THE PROBLEM OF THE UNITY OF THE SCIENCES: BACON TO KANT

THE
PROBLEM OF
THE UNITY
OF
THE SCIENCES:
BACON
TO KANT

ROBERT McRAE

Professor of Philosophy, University of Toronto

UNIVERSITY OF TORONTO PRESS

University of Toronto Press

Diamond Anniversary 1961

Preface

STRANGER. And here, if you agree, is a point for us to consider.
THEAETETUS. Namely?
STRANGER. The nature of the Different . . . appears to be parcelled out, in the same way as knowledge.
THEAETETUS. How so?
STRANGER. Knowledge also is surely one, but each part of it that commands a certain field is marked off and given a special name proper to itself. Hence language recognizes many arts and many forms of knowledge.[1]

THIS NOTION, to which passing reference is made in the *Sophist*, that all knowledge is in some sense one, has a history which begins with the Greeks, and of which the most spectacular recent phase is to be found in the series of International Congresses for the Unity of Science which took place in the period between the two world wars. In this study I am concerned with part only of this long history, that of the classical period of modern philosophy, the seventeenth and eighteenth centuries. At the beginning of the period, both Bacon and Descartes, the two philosophers most explicitly concerned with producing a complete revolution in the sciences, regarded unification as an essential element in their programmes, and the demand for it was made in the earliest of their philosophical writings. Leibniz, confronting a revolution that was already well advanced, was profoundly convinced that the sciences had now reached a unique, critical point at which a choice had to be made at once between a possible reversion to barbarism and a spectacular advance towards the perfection of civilization. The means for preventing the one and making possible the other was the logical integration of all knowledge in a "demonstrative encyclopaedia." This project, which he failed to achieve, was, as

[1]Plato, *Sophist*; *Plato's Theory of Knowledge*, tr. F. M. Cornford (London, 1935), 257 c.

Couturat has shown, a dominating preoccupation throughout his life.

It was primarily with a view to the advancement of the sciences and with the means to new discoveries that unification was sought by Bacon, Descartes, and Leibniz. But in the following century there was a significant change in attitude which has been recorded by Condorcet in his *Picture of the Progress of the Human Mind* (1795). "Soon," he says, "there was formed in Europe a class of men who were concerned less with the discovery or development of the truth than with its propagation, men who, whilst devoting themselves to the tracking down of prejudices in the hiding places where the priests, the schools, the governments and all long-established institutions had gathered and protected them, made it their life-work to destroy popular errors rather than to drive back the frontiers of human knowledge. . . ."[2] To achieve their ends they used "every mood from humour to pathos, every literary form from the vast erudite encyclopaedia to the novel or the broadsheet of the day."[3] As Condorcet implies, the chief concern of the French Encyclopaedists was with the social importance of the unification of knowledge. In their motives they were social reformers rather than scientific reformers. They lived in an age that accepted the apparent finality of Newton's achievement, and their eyes were less on the future of science than on the integration of already achieved results for the sake of public instruction and of the ushering in of a new era of universal enlightenment.

With Kant we find still another kind of concern with the unity of knowledge, sometimes referred to, with little sympathy, as his passion for architectonic. It has been said that Kant clings to architectonic "with the unreasoning affection which not infrequently attaches to a favourite hobby."[4] It would thus appear to rank with such foibles as the punctuality of his daily walk. But Kant would have regarded

[2] A.-N. de Condorcet, *Picture of the Progress of the Human Mind* (1795), tr. June Barraclough (New York, 1955), 136 f.

[3] *Ibid.*

[4] N. K. Smith, *A Commentary to Kant's 'Critique of Pure Reason'* (New York, 1950), xxii. Commenting on the chapter in the Transcendental Doctrine of Method entitled "The Architectonic of Pure Reason" Kemp Smith writes, "Adickes very justly remarks that 'this is a section after Kant's own heart, in which there is presented, almost unsought, the opportunity, which he elsewhere so frequently creates for himself, of indulging in his favourite hobby.' The section is of slight scientific importance, and is chiefly of interest for the light which it casts upon Kant's personality." *Ibid.*, 579. On the other hand, C. S. Peirce had immense admiration for this "splendid

this amusement over his concern with unity of system as directed against the very activity of philosophizing itself, for he believed that the ultimate task of the philosopher is to ascertain the systematic relation of all knowledge to the essential ends of human reason—a task of architectonic. To be a philosopher, as opposed to being a scientist concerned only with the logical perfection of his own particular science, is necessarily to seek the systematic unity of all the knowledge provided in the various sciences. The search for unity is not a philosopher's hobby, and Kant's views on the ideal of unity merit the same serious consideration that is accorded to other aspects of his thought.

In this study I am concerned with the unity of the sciences to the extent to which the determination of how this unity is to be conceived presented itself to philosophers as a specifically philosophical or logical problem. It is not, therefore, an essay in the history of ideas, in which the task of the historian would be to show the idea of unity at work in a multitude of different cultural contexts—in the founding of academies and scientific societies, the making of dictionaries and encyclopaedias, the devices for unity of language, and the educational projects such as those of Comenius and the Pansophists or the encyclopaedic curriculum of eighteenth-century German universities. If any of these are mentioned it is only as they bear directly on the different philosophical theories of how the unity of the sciences is to be conceived.

Nor is this a history of the classification of the sciences, although any classification involves some theory of the way in which the different kinds of knowledge are related to one another and thus might be said to attribute to them a certain kind of unity. But a review of such classifications shows that the purposes behind them can be totally opposed to one another. Descartes and Leibniz classify only for the purpose of uniting; Spinoza, Malebranche, Locke, or Berkeley do so in order to separate. If we take a classification which in its general outlines is common to both Bacon and the French

---

third chapter of the Methodology." "The universally and justly lauded parallel which Kant draws between a philosophical doctrine and a piece of architecture has excellencies which the beginner in philosophy might overlook; and not least of these is its recognition of the cosmic character of philosophy." *Philosophical Writings of Peirce*, ed. Justus Buchler (New York, 1955), 72.

Encyclopaedists we find that with Bacon the classification is *not* designed to exhibit the unity of the sciences, while for Diderot that is its only role. When Bacon speaks of the unity of the sciences he is referring to something completely different from what is revealed in such a scheme, and he makes his classification only with the warning that it must not be taken as dividing the sciences from one another. Consequently I have not included an account of Bacon's classification, but I have had to pay attention to Diderot's, even though it is for the most part a borrowed one, and, not being original, is much less interesting than Bacon's.

Finally, there is nothing in this study which refers to such philosophers as Hobbes or Hume, who might well be said to have ascribed unity to the sciences, Hobbes in ordering them in a single deductive system resting on definitions, and Hume in relating them to one central science, the science of man.[5] There is no indication, however, that either philosopher was specifically concerned with unity of knowledge as a problem, or attached any significance to it. They may suppose a unity, but they do not discuss it. In short, this is not a study of philosophical presuppositions, but only of what is directly said on the subject of unity by philosophers.

The differences in the conceptions of unity are in some cases so great that one may wonder indeed whether they belong within the discussion of the same subject. This has made it necessary to bring them together at the outset, systematically and unhistorically, to make clear that there is in fact a common subject. Only after that does it make sense to give to each of them in separate chapters the fuller treatment which they require. These later chapters sometimes contain sections considerably amplifying the material brought together in the first. References have, therefore, been provided in the footnotes to those places where the fuller treatment is to be found. In some cases this involves an unavoidable repetition, but it will be a repetition within a different but no less necessary context.

I generally use the words "science" and "philosophy" after the manner of the authors whose theories are examined. The use varies. Although for the most part in the seventeenth and eighteenth centuries the two expressions are equivalent, this is not always the case.

[5]Hobbes, *De Corpore* (London, 1655), chapter vi; Hume, Introduction, *A Treatise of Human Nature* (London, 1739–40).

The context, however, usually makes quite clear any differences there are in the senses of the words for the particular occasions in which they are being used.[6] The word "sciences" in the title of this work carries its wider traditional meaning in that it refers to all the disciplines which yield knowledge.

I owe special acknowledgment to two classic studies of philosophers whose views are considered in this book. One is F. H. Anderson's *Philosophy of Francis Bacon* in which for the first time Bacon's philosophical materialism received full attention and emphasis, particularly in chapter IV, "Bacon's Revival of Materialism: His Interpretation of Fables," and chapter V, "Bacon's Materialism: Atoms and Motion." I found this materialism to be fundamental for the understanding of Bacon's conception of the unity of the sciences. The other work is Louis Couturat's *La Logique de Leibniz* with its thorough study, both historical and critical, of the Universal Characteristic, the Encyclopaedia, and the General Science. I am grateful to Professor T. A. Goudge, Professor C. W. Webb, and Professor C. D. Rouillard, and to the publisher's anonymous reader for their generosity in making criticisms and suggestions on numerous points;

---

[6]Hobbes asserts the identity of the terms in common usage when he refers to "Science, that is knowledge of consequences, which is called also philosophy." *Leviathan* (London, 1651), chapter IX. A hundred years later Diderot remarks that the words are synonyms when he gives as one of the three main branches of knowledge, "Philosophy or Science." *Prospectus, Œuvres complètes de Diderot*, ed. J. Assézat (Paris, 1875–7). A single writer, however, can at one time identify them and at another distinguish them. Thus Locke speaks of "*Physica* or Natural Philosophy" as one of the three "Divisions of the Sciences," *Essay Concerning Human Understanding* (London, 1690), IV, xxi, while earlier under the heading "Hence no science of bodies," he expresses doubt that "experimental philosophy in physical things" will ever be "scientifical." *Ibid.*, iii. In the context of the former statement "science" is equivalent to "knowledge," in that of the latter it refers to what is capable of demonstrative certainty. Bacon, while often enough using the terms "philosophy" and "science" interchangeably, can also speak of "Natural Science or Theory" as being merely a part of "Natural Philosophy." Here the word "science" is reserved for the speculative part, or the inquiry into causes, in contrast to the operative part of natural philosophy, which has to do with the production of effects. When Descartes excludes the science of mathematics from philosophy (*The Philosophical Works of Descartes*, tr. E. S. Haldane and G. R. T. Ross [Cambridge, I, 1931; II, 1934], I, 91, 211) and Kant contrasts them (*Critique of Pure Reason* [I, 1781; II, 1787], tr. N. K. Smith [London, 1933], A 713 = B 741 ff.) it is evident that they give science a wider extension than philosophy, for both regard philosophy as science also. For a discussion of the history of the word "science" from the time of Bacon on, and of the differences in English, French, and German usage, see J. T. Merz, *A History of European Thought in the Nineteenth Century* (Edinburgh and London, 1896), I, 89 ff., 168 ff.

to the Publications Fund of the University of Toronto Press and to the Humanities Research Council of Canada (out of funds provided by the Canada Council) for grants to make publication possible; to the editor of *Philosophy and Phenomenological Research* for his permission to include as chapter VII, with minor alterations, an article of mine on Kant which was published in that journal in September 1957; and to Miss Eunice Lamb who so kindly spent hours unravelling my handwriting to make a manuscript.

R.F.McR.

Contents

Contents

THE PROBLEM OF THE UNITY OF THE SCIENCES: BACON TO KANT

I. THE IDEAL OF UNITY

THE "UNITY OF THE SCIENCES" is not a single, simple conception, as is shown by the range of opinion about it, not only in the international movement of the twentieth century, but also in the classical period of modern science, which is our concern here. An individual established science, like any one of the mathematical sciences or of the various branches of physics, actually possesses a high degree of integration. For the philosopher to determine what constitutes its unity is only to make explicit something which is, if not fully, at least partially, present in it. In spite of the fact that there are, as we shall see later, different views as to what makes this unity, the problem remains essentially one of clarification. On the other hand, the systematic unity of all the sciences taken together is, as Kant observes, "a mere idea of a possible science which nowhere exists *in concreto* . . . where is it, who is in possession of it, and how shall we recognize it?"[1]

It was Kant's view that the demand for the unity of the sciences was contained in the very nature of philosophy, and that we are compelled to seek it, even if we are not at all clear as to precisely what it is. If, however, we do not know what this ideal is, it is plain that we shall not even know if the unity of the sciences is a possible conception. If it is remains an open question whose importance, despite all scepticism, is indicated in the institution in our own times of the Tarner Lectures in Cambridge. The subject of these lectures is: "The Philosophy of the Sciences and the Relations or Want of Relations between the Different Departments of Knowledge." When Professor Whitehead gave the first of these—in *The Concept of Nature*—he referred, like Kant, to the compulsiveness of the ideal of unity, without, however, committing himself to the view that the different kinds of

[1]*Critique of Pure Reason* (I, 1781; II, 1787), tr. N. K. Smith (London, 1933), A 838 = B 866.

knowledge could actually possess it. But he maintained that even if it could be proved that the different sciences are without relation to one another, this disproof would itself constitute important knowledge for us. "You cannot," he said, "abandon the later clause of the definition; namely that referring to the relations between the sciences, without abandoning the explicit reference to an ideal in the absence of which philosophy must languish from lack of intrinsic interest. . . . That far-off ideal is the motive power of philosophic research; and claims allegiance even as you expel it."[2] It may well be an *ignis fatuus* leading the philosopher on. Where did it come from?

THE UNITY OF A LOGICAL SYSTEM WITHOUT DIVERSITY OF SCIENCES

It is, undoubtedly, the conception of unity in *a* science which gives rise to one conception of how the unity of all knowledge is possible. Descartes explains in the *Discourse on Method* (1637) how the idea of this unity arose with him as a result of considering the nature of geometry. "Those long chains of reasoning," he wrote, "simple and easy as they are, of which geometricans make use in order to arrive at the most difficult demonstrations, had caused me to imagine that all those things which fall under the cognizance of man might very likely be mutually related in the same fashion."[3] The individual science of geometry provided him with the paradigm for the unification of all knowledge. But certain problems at once arise when we attempt to conceive the unity of the sciences as identical in nature with the unity which characterizes any one of the particular sciences. As an initial and minimum description of the unity of *a* science, it may seem safe to say that there must be two elements present. In the first place it might be said that a science is one in virtue of possessing one subject-matter which is peculiarly its own, so that there is some knowledge which can be said to belong within that science, and other knowledge which is excluded. It may not always be easy to state exactly what the subject-matter of the science is, but the experienced scientist will have a fairly clear sense of the sort of questions which are improper to his science. Secondly, it might be said that the variety

[2]A. N. Whitehead, *The Concept of Nature* (Cambridge, 1920), 1 f.
[3]*Discourse on Method* (1637), II, *The Philosophical Works of Descartes*, tr. E. S. Haldane and G. R. T. Ross (Cambridge, I, 1931; II, 1934), I, 92.

of knowledge contained in the science will be unified in a certain logical structure. A science cannot be merely a collection of propositions bearing on the same subject-matter; they will have to be logically interconnected before we will consider them as together forming one science.

If there is one general science which brings *all* knowledge into the same kind of unity as that found in any one of the particular sciences, then, in the light of what has just been said, the question would have to be asked of this science, What is its subject-matter? We already possess certain groups of sciences, the mathematical sciences, the natural sciences, the social sciences, whose members form a class because they share a common subject-matter. The natural sciences might, for example, be said to have nature as their subject-matter. At the same time this does not necessarily mean that taken all together the members of the class form one science. To make them one science would seem to require the taking of one of their members as basic and reducing all the rest to it. If in turn the different classes of sciences are to be unified into a single science, their members too would all have to be reduced to the one privileged member of the one privileged class. For example, in a paper contributed to the fifth International Congress for the Unity of Science in 1939, Professor Feigl asserts that "mechanics, astronomy, acoustics, thermodynamics, optics, electricity, magnetics, and chemistry are integrally united (at least to an astounding extent) in the theories of relativity and quanta."[4] It is his hope that just as the various sciences within this group attain unity by reduction, so also biology, psychology, and the social sciences will ultimately attain unity by reduction to physical theory.

Besides the method of reduction, unity of subject-matter can be secured also by the rigid exclusion of everything which does not come under one class of subject-matter and by the assertion that it alone is fit and appropriate for scientific inquiry. Such a restriction upon the possible scope of science might be determined in a number of ways; for example, by a particular epistemological theory, or by a theory of the nature of scientific method, or by logical analysis and theory of meaning. At the beginning of the seventeenth century we

[4]Herbert Feigl, "Unity of Science and Unitary Science," *Readings in the Philosophy of Science,* ed. Herbert Feigl and May Brodbeck (New York, 1953), 383.

find such an exclusion determined, in the case of Francis Bacon, by
the edicts of revealed religion, which forbids scientific investigation of
the nature of God, or of the nature of man's rational soul, which is
the image of God, or any questions concerning the relation of this
soul to the body, or any attempt to find the grounds of the moral law.
Nature alone is left as the legitimate subject of science. As a consequence, natural philosophy or science does not have for Bacon the
status of being one among the particular sciences. It is, on the contrary,
designated as "the mother" of the particular sciences. "Let no man,"
he says, "look for much progress in the sciences . . . unless natural
philosophy be carried on and applied to particular sciences, and particular sciences be carried back again to natural philosophy." This is
true not only for such sciences as "astronomy, optics, music, a number
of mechanical arts, medicine itself," but also for "what one might
more wonder at, moral and political philosophy, and the logical
sciences."[5]

Let us suppose, however, that the subjects remain irreducibly
various, and that they do not fall under one class. Is it then any
longer possible to consider them as susceptible of integration within
one all-comprehensive science? To answer that it is possible would
seem to involve the denial of our original supposition that unity of
subject-matter is a necessary condition of the unity of a science. And
this supposition was denied by Descartes.

In a letter to Mersenne, Descartes writes,

It is to be observed in everything I write that I do not follow the order of
subject matters, but only that of reasons, that is to say, I do not undertake
to say in one and the same place everything which belongs to a subject,
because it would be impossible for me to prove it satisfactorily, there being
some reasons which have to be drawn from much remoter sources than
others; but in reasoning by order, *a facilioribus ad difficiliora*, I deduce
thereby what I can, sometimes for one matter, sometimes for another,
which is in my view the true way of finding and explaining the truth;
and as for the ordering of subject matters, it is good only for those for
whom all reasons are detached, and who can say as much about one difficulty as about another.[6]

[5]*Novum organum* (1620), I, lxxx, *The Works of Francis Bacon*, ed. Ellis, Spedding, and Heath (Boston, 1864), VIII, 112. See further chapter II, 29 ff., of the
present book.

[6]*Letter to Mersenne*, 24 Dec., 1640, *Descartes, Correspondance*, ed. C. Adam and
G. Milhaud (Paris, 1936 *et seq.*), IV, 239. See further chapter III, 56ff., of the
present book.

In other words, if we order our knowledge according to subject-matters, we shall sacrifice the essential element which makes knowledge scientific, the logical interconnection of parts, as this is typified in the science of geometry. Leibniz was to point out later that in geometry itself the demonstrative order does not permit everything which belongs to the same subject to be dealt with in the same place. Descartes' conception of the unity of the sciences has the important implication that there cannot, strictly speaking, be any classification of the sciences in the sense of a division. To isolate a science according to subject-matter is to deprive it of its scientific character and to dissolve it into a mere collection of detached truths. Hence there can be only one science, a universal one. It is undefinable by its subject-matter, and its parts are not distributed by subject-matters. In the opening passage of his *Regulae*, Descartes deplores the division of labour in science. Because men have seen the necessity for such a division in the exercise of the manual arts, "they have held the same to be true of the sciences also, and distinguishing them according to their subject-matter, they have imagined that they ought to be studied separately, each in isolation from the rest. But this is certainly wrong."[7] And one reason why it is wrong, as he indicates in his letter to Mersenne, is that the scientific ordering of truths cuts right across their ordering by subject-matter.

Leibniz, too, shared this view that there is only one science, and that the distribution of knowledge under individual sciences is not based on any logical and therefore necessary unity in the parts of these sciences, but is entirely arbitrary, and instituted only for convenience. And in so far as these sciences are distinguished by their subject-matters, they will always, he said, be encroaching upon each other's rights and be continually at war. "It is usually found that one and the same truth may be put in different places according to the terms it contains, and also according to the mediate terms or causes upon which it depends, and according to the inferences and results it may have. A simple categoric proposition has only two terms; but a hypothetic proposition may have four, not to speak of complex statements."[8]

[7]*Regulae ad directionem ingenii* (1628), I, H. R., I, 1.
[8]*New Essays Concerning Human Understanding* (1704), IV, xxi, 4, tr. A. G. Langley (La Salle, 1949), 623.

To organize knowledge logically is, for Leibniz, to produce but one science, which can be presented either in the synthetic or theoretic order, i.e., in the order of proofs, or in the analytic or practical order, that of discovery. If we should now want to know everything in this one science which could be ranked under any one subject-matter, we should need an index to guide us—an index "either systematic, arranging the terms according to certain predicaments which would be common to all the notions, or alphabetical according to the languages received among scholars. Now this index would be necessary in order to find together all the propositions into which the term enters in a sufficiently remarkable manner; for according to the two preceding ways, when the truths are arranged according to their origin or use, truths concerning one and the same term cannot be found together. . . . The index may and should indicate the places where are found the important propositions which concern one and the same subject."[9]

When the French Encyclopaedists found it necessary to introduce some system of the classification of the sciences in order to produce their *Encyclopaedia,* they insisted on the absolute arbitrariness of all such divisions. Diderot and d'Alembert could well be included with the old Nominalists, who, said Leibniz, "believed that there were as many particular sciences as truths, which they composed after the wholes, according as they arranged them."[10] The logical filiations of the different parts of knowledge were, for Diderot and d'Alembert, entirely independent of any arbitrary grouping by subject-matter, and hence, like Leibniz, they required some device for reconciling these two distinct types of order—the scientific and the classificatory. Because it is in these logical filiations that the unity of the arts and sciences is revealed, an elaborate network of cross-references became necessary among the contents of their alphabetically listed subjects. The use of these cross-references was, said Diderot, "the most important part of our encyclopaedic scheme." It was by means of them that what otherwise would merely have been a dictionary of the arts and sciences was to be transformed into an encyclopaedia. The most significant difference between this encyclopaedia and Leibniz's projected demonstrative encyclopaedia was that where Leibniz made

---

[9]*Ibid.,* 625. See further chapter IV, 83 ff., of the present book.
[10]*Ibid.,* 623.

the logical ordering of knowledge primary, and introduced the index
to provide classification by subject, Diderot and d'Alembert made
classification by subject primary, and introduced cross-references to
establish the logical connections.[11] The reason for this difference lay
in their profound scepticism of the extent to which the logical integration of knowledge was capable of being realized. A complete
demonstrative system of knowledge, such as Leibniz envisaged, was
entirely out of the question. Theirs was to be the unity of an imperfect
network, not of a logical system.

THE UNITY OF A LOGICAL SYSTEM WITH DIVERSITY OF SCIENCES

Kant's theory of the unity of science implies a radical criticism of
these views of Descartes, Leibniz, and the French Encyclopaedists.
Descartes had condemned the distribution of knowledge in different
sciences on analogy with the division of labour in the arts. Kant
supported it. "All trades, arts, and handiworks have gained by division
of labour, namely, when, instead of one man doing everything, each
confines himself to a certain kind of work distinct from others in the
treatment it requires, so as to be able to perform it with greater
facility and in the greatest perfection. Where the different kinds of
work are not so distinguished and divided, where everyone is jack-of-
all trades, there manufactures remain still in the greatest barbarism."[12]
And the same, he maintains, is true for the sciences. Some of Kant's
strongest expressions of contempt in his writings (e.g., he speaks of
"bunglers" and "disgusting medley") are reserved for those who fail
strictly to isolate the different kinds of scientific inquiry. "We do not,"
he says, "enlarge but disfigure sciences, if we allow them to trespass
upon one another's territory."[13]

"Every science is a system in its own right . . . we must . . . set to
work architectonically with it as a separate and independent building.
We must treat it as a self-subsisting whole."[14] This statement puts
the conflict of Kant's views with those of Descartes and Leibniz in

[11]See further chapter VI, 121 f., of the present book.
[12]*Fundamental Principles of the Metaphysic of Ethics* (1785), Preface, tr. T. K.
Abbott (London, 1946), 2 f.
[13]*Critique of Pure Reason*, B ix.
[14]*Critique of Teleological Judgement* (1790), I, 7, tr. J. C. Meredith (Oxford,
1928), 31.

the strongest light. Kant is implying that there is a direct connection between the classificatory and the logical ordering of knowledge—each science is a self-contained logical system; and there is nothing arbitrary about the divisions of the sciences—each is "a system in its own right." At the root of this conflict we find two radically opposed conceptions of the nature of a logical system. For convenience we can, taking the terms from our authors themselves, label them respectively the "combination" theory and the "organic" theory.

If we look at our knowledge, says Descartes, with a view to determining the way in which some truths are dependent on others, we shall have to distinguish those things which are known *per se* and those things which are deduced from them. The first are characterized by "the extremest simplicity," and he calls them simple essences or natures. "We must note," he says, "that there are but few pure and simple essences which either our experiences or some sort of innate light in us enable us to hold as primary and existing *per se*, not as depending on any others."[15] The rest of our knowledge is of "the complex and composite," and can consist in nothing but the combination of these atomic simple natures. "No knowledge is at any time possible of anything beyond those simple natures and what may be called their intermixture or combination with one another."[16] Deduction is an art of combination. The same conception of a repertory of logical simples is found in Leibniz, and the same view that all the mind can do in extending its knowledge is to combine. "The fruit of several analyses of different particulars will be the catalogue of simple thoughts, or those which are not very far from being simple. Having the catalogue of simple thoughts, we shall be ready to begin again *a priori* to explain the origin of things starting from their source in a perfect order and from a combination or synthesis which is absolutely complete. And that is all our soul can do in its present state."[17] The logic of invention is nothing but an application of the *ars combinatoria*.

Where, however, for Descartes and Leibniz a logical system, considered as a *whole*, is merely a function of its simple constituent elements, for Kant this whole is an organic one; that is to say, the

[15]*Reg.* VI, H. R., I, 16.
[16]*Ibid.*, XII, H. R., I, 43. See further chapter III, 55 f., of the present book.
[17]*On Wisdom* (ca. 1693), *Leibniz, Selections,* ed. P. P. Wiener (New York, 1951), 80. See further chapter IV, 76 f., of the present book.

conception of the whole is logically prior to that of its parts; the whole determines what those parts shall be and how they shall be related to one another. Every science in so far as it constitutes a *system* rests on what Kant, in a very technical sense, calls an antecedent "idea."

This idea is the concept provided by reason—of the form of a whole—in so far as the concept determines *a priori* not only the scope of its manifold content, but also the positions which the parts occupy relatively to one another. The scientific concept of reason contains, therefore, the end and the form of that whole which is congruent with this requirement. The unity of the end to which all the parts relate and in the idea of which they all stand in relation to one another, makes it possible for us to determine from our knowledge of the other parts whether any part be missing, and to prevent any arbitrary addition, or in respect of its completeness any indeterminateness that does not conform to the limits which are thus determined *a priori*. The whole is thus an organized unity (*articulatio*), and not an aggregate (*coacervatio*). It may grow from within (*per intussusceptionem*), but not by external addition (*per appositionem*). It is thus like an animal body, the growth of which is not by the addition of a new member, but by the rendering of each member, without change of proportion, stronger and more effective for its purposes.[18]

Kant employs this conception of the system of a science as an organic whole not only to account for the divisions between the sciences, but also to account for their unity. "Not only is each system articulated in accordance with an idea," he says, "but they are one and all organically united in a system of human knowledge, as members of one whole, and so as admitting of an architectonic of all human knowledge, which, at the present time, in view of the great amount of material that has been collected, or which can be obtained from the ruins of ancient systems, is not only possible, but would not indeed be difficult."[19] Just as each science rests on an idea, so the systematic unity of all the different sciences in one whole must rest on an idea. The totality of the sciences is an organism comprising lesser organisms.

There is, as we said, nothing arbitrary for Kant in the classification of the sciences. There is only one such classification which is objectively valid. Where Leibniz admits classification only in terms of convenience, or of optionally entertained purposes which are external

---

[18]*Critique of Pure Reason*, A 833 = B 861. See further chapter VII, 135 f., of the present book.

[19]*Ibid.*, A 835 = B 863.

to the mind's logical function of combining concepts, Kant regards these purposes as arising within the exercise of the purely logical function of the mind.

Kant's analysis of the syllogism is designed to show that reason in exercising its purely logical function has an end in view, namely, to reduce the manifold knowledge provided by the understanding to the highest degree of unity.[20] But reason cannot conceive this unity at which it necessarily aims except by giving the idea of it an object. The objects of these ideas which serve to distinguish the different sciences are not, however, the subject-matters of the science, for even if reason is compelled to postulate them there is no objective ground for asserting that such objects exist. Their sole function is to guide reason in its attempt to connect the knowledge provided by the understanding into scientific knowledge, that is to say, to give it systematic unity. Reason follows these ideas "only as it were asymptotically, i.e. ever more closely without ever reaching them."[21]

The basis for the divisions of the sciences does not, then, for Kant, lie in the differences of their subject-matters, but in their ideas.[22] And these ideas have their origin in reason as exercising its purely logical function. In one case Kant provides us with a detailed working out of this principle of division. It is to be found in relation to his basic classification of the natural sciences. These are three—physics, psychology, and a third science which is a systematic union of physics and psychology. What is the foundation of this classification? It is to be found in the three ways in which our thought is related in judgments. There is, first, the relation of the predicate to the subject in the categorical judgment; secondly, the relation of the ground to its consequence in the hypothetical judgment; and, third, the

---

[20]See chapter vii, 132, of the present book.

[21]*Critique of Pure Reason*, A 663 = B 691.

[22]Kant grants that some sciences can be distinguished simply by their difference of subject-matter, but this will not hold universally (*Prolegomena zu einer jeden künftigen Metaphysik, die als Wissenschaft wird auftreten können* [Riga, 1783], § 1). It will not, for example, serve to distinguish philosophy from mathematics, "for philosophy applies to everything, and therefore also to *quanta*, and mathematics applies in part to everything, inasmuch as everything has quantity. It is only the *different kind of rational knowledge, or of the use of reason* in mathematics and philosophy that constitutes the specific difference between these two sciences. Philosophy is *rational knowledge from mere concepts*; mathematics, on the contrary, is *rational knowledge from the construction of concepts.*" *Introduction to Logic*, III, tr. T. K. Abbott (London, 1885), 13. If a science differs from others by its subject-matter it will be because this subject-matter like any other feature peculiar to it is determined by the "idea" of the science.

relation of logical opposition in the disjunctive judgment in which the constituent propositions exclude one another, but at the same time, when taken together, are exhaustive of the whole of the knowledge in question. There are also three kinds of syllogism corresponding to these three ways in which our thought is related in judgments. And accordingly, when reason seeks, as it necessarily must, the totality of conditions for any given conditioned knowledge, it will advance through a series of prosyllogisms towards the unconditioned. By the categorical syllogism it will seek "the subject which is never itself a predicate"; by the hypothetical syllogism it will seek "the presupposition which itself presupposes nothing"; and by the disjunctive syllogism, "such an aggregate of the numbers of the division of a concept as requires nothing further to complete the division."[23] In this way reason will be compelled to postulate three ideas of the unconditioned to direct its logical activity: it will postulate the unconditioned unity of the thinking subject to which all our representations are related, the absolute unity of the series of conditions for 'these representations considered as appearances of objects, and thirdly, the absolute unity of the conditions of all objects of thought in general. In so far as any of these three is considered to be the object of a science, that science is a pseudo-science, for these objects lie beyond all possibility of being known. The proper function of the ideas is purely to regulate scientific inquiry, and this function they perform as ideas of "the form of a whole of knowledge—a whole which is prior to the determinate knowledge of the parts and which contains the conditions that determine *a priori* for every part its position and relation to the other parts."[24] It follows, then, that there are three such wholes, answering to each of the three ideas, and these will constitute the three natural sciences of Kant's classification. The idea will in each case "determine the proper content, the articulation (systematic unity), and limits of the science."[25]

THE UNITY OF A TELEOLOGICAL SYSTEM

Up to now it has been the unity of the sciences as conceived by analogy with the unity of the individual science, which has been considered, beginning with the notion that the individual science is

[23]*Critique of Pure Reason*, A 323 = B 380.
[24]*Ibid.*, A 645 = B 673.
[25]*Ibid.*, A 834 = B 862. See further chapter VII, 137 f., of the present book.

unified by its subject-matter and by the logical interconnection of its parts. It was seen that with Descartes and Leibniz unity by subject-matter is abandoned as being incompatible with unity by logical integration. The consequence of this is that there is, for them, strictly speaking, no plurality of individual sciences. There is only one universal science, containing within it all scientific knowledge. With Kant, on the other hand, there is a plurality of individual sciences, and their identities are fully maintained in that one ultimate systematic whole to which they all belong. But Kant, too, abandons the notion of unity by subject-matter. In its place he introduces the regulative "idea" of the science. It is this which holds together the many different concepts which belong within the science. And it is similarly a regulative "idea" which provides the basis for the unity of all the sciences in one system of knowledge.

In contrast with these views there is a conception of the systematic connection of all the sciences which is not based on analogy with the unity of the individual science. Such a conception can be found in Condillac. An individual science is one, for Condillac, by virtue of being merely the progressive analysis of a single idea by means of language. For an idea to be grasped clearly it must be articulated in propositions. The function of the proposition is to break down the idea into components, or analyse it. A proposition, therefore, asserts nothing but an identity. The entire system of propositions which comprises a science consists in a succession of translations of an original proposition asserting such an identity. Thus an individual science regarded as a logical system is only the progressive linguistic articulation of a single idea. For example, the whole of mathematics is contained in the idea of the single word "to measure." All the truths of mathematics are different verbal expressions of an initial definition of this word.[26]

So much for the individual science as a system. Condillac also regards all the sciences taken together as comprising a system. His conception of this system, however, is not taken over from that of the individual science. If it were, it would mean asserting that there is a single idea containing within it the ideas of all the individual sciences and these sciences taken collectively would be the linguistic expression of this one idea. But Condillac never makes any such proposal.

[26]See chapter v, 100 f., of the present book.

Rather, he turns to a consideration of the purposes of the different sciences in order to find there the basis for their integration in a single system. It is these purposes which, as we shall see, determine the fundamental classification of the sciences. For Condillac man is related to the world in which he lives as an organism is related to its environment. All his needs and their modes of satisfaction are a function of this relation, and hence they too will all be organically coordinated with one another or form a system. It is need alone which provides the motive for human inquiries. Corresponding to each need there will be the idea of the thing necessary for its satisfaction, and the analysis of this idea gives rise to a whole series of constituent ideas. Just as our needs form a system, so also our ideas will form a corresponding system. And because each of the different sciences is merely the analysis of a fundamental idea answering to a fundamental need, the different sciences will form an organic system. This implies for Condillac that the constituents of the different fundamental ideas which are revealed by analysis will not be confined in their membership to only one of the sciences, but may be common to several, "for the same objects and consequently the same ideas are often related to different needs."[27] It may be noted, finally, that with Condillac the unification of the sciences does not, as with Descartes and Leibniz, involve the obliteration of their divisions. Needs may be organically interrelated, but they are nevertheless distinguishable. At the same time he considered the passion for separating out different intellectual disciplines a mere relic of scholasticism and had little patience with it. Because the sciences are organically interrelated they shed light upon one another, and they should all be studied together.

Condillac and Kant make strange company. But Kant too conceived that the purposes of the sciences bring them into a systematic unity which he believed to be the special concern of the philosophers to establish. For Kant the scientist himself is not and should not be concerned with any external purpose to be served by his science. He should be concerned only with achieving its greatest logical perfection. "Our understanding, moreover, is so constituted that it finds satisfaction in mere insight, and even more than in its resulting

[27]*Cours d'études, De l'art de penser* (1775), V, *Œuvres philosophiques de Condillac*, ed. Georges Le Roy (Paris, 1947–51), I, 726b. See further chapter v, 104 f., of the present book.

utility."[28] Nevertheless the philosopher must in the end ask, What does this purely speculative knowledge contribute to the ultimate end of human reason? At this point philosophy becomes a teleology of human reason.

The mathematician, the natural philosopher, and the logician, however successful the two former may have been in their advances in the field of rational knowledge, and the two latter more especially in philosophical knowledge, are yet only artificers in the field of reason. There is a teacher, [conceived] in the ideal, who sets them their tasks, and employs them as instruments, to further the essential ends of reason. Him alone we must call philosopher; but as he nowhere exists, while the idea of his legislation is to be found in that reason with which every human being is endowed, we shall keep entirely to the latter, determining more precisely what philosophy prescribes as regards systematic unity . . . from the standpoint of its essential ends.

Essential ends are not as such the highest ends; in view of the demand of reason for complete systematic unity, only one of them can be so described. Essential ends are therefore either the ultimate end or subordinate ends which are necessarily connected with the former as means. The former is no other than the whole vocation of man. . . .[29]

As each of the sciences serves some essential end of human reason, they will all be systematically unified in relation to that science which is concerned with the ultimate end of human reason—moral philosophy or, more specifically, moral theology.[30]

We have now reviewed the use of one conception of the unity of the sciences, namely unity of system, and we have found two kinds of system, the logical system (either combinatory or organic) and the teleological system, though the logical and the teleological are integrally united for Kant—for even in the exercise of its purely logical function reason is directed by an end. In the teleological system the sciences are unified through the organization of their ends in relation to one common or supreme end. For Condillac this common end is the preservation and well-being of the human organism in relation to its environment. For Kant it is "the whole vocation of man."

UNITY OF METHOD

A second principal conception of the unity of the sciences in the seventeenth and eighteenth centuries is that of unity of method.

[28]*Introduction to Logic* (1800), tr. T. K. Abbott (London, 1885), 33.
[29]*Critique of Pure Reason,* A 840 = B 868.
[30]See chapter VII, 139 f., of the present book.

This unity was asserted by all the philosophers whom we have been considering with the exception of Kant. Let us look first at Kant's denial of it.

Kant distinguishes in the *Critique of Pure Reason* two ways in which logic can be treated. It can be concerned with the necessary rules of thought which are the condition for any use of the understanding at all, and without any regard to differences in the kinds of objects to which thought is directed. Or it can be concerned with the special use of the understanding, giving the rules of correct thinking with regard to certain *kinds* of objects, in which case it is an organon of this or that science. In his *Introduction to Logic*, however, Kant removes the latter kind of rules from logic altogether. Logic is not an organon of the sciences. It is not concerned with showing how some particular kind of knowledge is to be acquired. In order to know the rules of method applicable to a particular science you must already possess a fairly complete knowledge of the objects with which the science in question is concerned, and of the source of this knowledge. But logic itself "cannot meddle with the sciences, or anticipate their matter.[31] "Logic is not a general art of discovery, nor an organon of truth; it is not an algebra by means of which hidden truths may be discovered."[32] Logic proper had, in Kant's view, been brought to completion as a science in Aristotle's Analytic, though he makes the observation that Aristotle himself had treated it as an organon.

In direct contrast with this, Bacon, Descartes, Leibniz, and Condillac all identify scientific method, or what Kant calls "the organon of the sciences," with logic proper. Indeed, what Kant considered to be logic, namely syllogistic, is condemned by Bacon as a mere art of disputation, and according to Descartes it should be "transferred from Philosophy to Rhetoric."[33]

To identify the method of the sciences with logic is to claim for it the same universality of application that admittedly belongs to logic. It is to assert that its rules will operate without any regard to differences in the objects to which the understanding may be directed. Method will be the same whether the science is mathematics, physics, metaphysics, ethics, jurisprudence, or any other of the sciences

[31]*Introduction to Logic,* 3.
[32]*Ibid.,* 10.
[33]*Reg.,* X., H. R., I, 33.

enumerated by these philosophers. But can scientific method be formulated in total disregard of the nature of its objects? After all, it might be said, the universal methods of Descartes, Leibniz, and Condillac are merely generalizations of the method of mathematics— a method already worked out in relation to a specific kind of object. As Kant remarks, scientific method is commonly taught as a propae- deutic to the sciences, but if we consider the actual order in which our knowledge is acquired, this method is arrived at last of all when the science in question has already been well established. If this is correct, then it would only be possible for Descartes and Leibniz to arrive at their theories of method by reflection on the procedures of some established science, and for both of them, it might be said, that science was mathematics. What is required, however, is some justifica- tion for transferring this method from the objects of mathematics to all objects. Because the expression "generalization of the method of mathematics" contains the suggestion of an uncritical assumption of the possibility of this transference, it is important to consider the basis of these claims to universality of method.

Leibniz has explained how he first formed the idea of his *ars combinatoria*. It did not have its origin in reflection upon the method of mathematics, for at the time he did not yet have any acquaintance with mathematics, but in reflection upon the logic of Aristotle, Leib- niz recounts that among the questions he had raised as a student to his instructors in logic was

. . . a doubt concerning the categories. I said that, just as we have cate- gories or classes of simple concepts, we ought also to have a new class of categories in which propositions, or complex terms themselves, may be arranged in their natural order. For I had not dreamed of demonstrations at that time, and did not know that the geometricians were doing exactly what I was seeking when they arranged propositions in an order such that one is demonstrated from the other. . . . Upon making the effort to study this more intently, I necessarily arrived at this remarkable thought, namely, that a kind of alphabet of human thoughts can be worked out and that everything can be discovered and judged by a comparison of the letters of this alphabet and an analysis of the words made from them.[34]

This is the fundamental idea underlying his universal method, and it was to remain fundamental throughout all its subsequent develop-

[34]*On the General Characteristic* (ca. 1679), *Gottfried Wilhelm Leibniz, Philo- sophical Papers and Letters*, ed. L. E. Loemker (Chicago, 1956), 341. See further chapter IV, 77, of the present book.

ment as the Universal Characteristic. The method is as universal as the logic of Aristotle, and was simply regarded by Leibniz as that logic brought to its perfection. When later he acquired a knowledge of the mathematical sciences he discovered in them a highly successful exemplification of what he was seeking. "No more beautiful example of the art of combinations can," he says, "be found anywhere than in algebra"; and, he adds, "therefore, he who masters algebra will the more easily establish the general science of combinations, because it is always easier to arrive at a general science *a posteriori* from particular instances than *a priori*."[35] And Leibniz admits the extent to which he himself was dependent upon this *a posteriori* method of establishing the general science of combinations. He could hardly, he says, have attained it without the aid of mathematics.[36] It is, however, the antecedent recognition that mathematics is in fact a particular instance of the application of the general science, which makes possible for him the study *a posteriori* of the general science by means of mathematics.

Where Leibniz derives his conception of an alphabet of human thoughts directly from the traditional logic, and independently of mathematics, Descartes derives his from his theory of the nature of cognition. When investigating "the nature and scope of human knowledge"—and it is from such an investigation, he says, that "the whole method of inquiry comes to light"[37]—the objects of knowledge are not to be considered as they are "in their more real nature," but "only in relation to our understanding's awareness of them." From this point of view their nature is entirely determined by the nature of the mind in the exercise of its function of knowing. In knowing there is only one thing the mind can do. It is capable only of a purely passive perception, or mental vision or intuition, which remains exactly the same, whatever its objects. To know is to "grasp each fact by an act of thought that is similar, single, and distinct."[38] The objects which are alone appropriate to this solitary cognitive function of the mind are simple natures or essences, and after that such connections between them as can be grasped in a single mental vision.

[35]*Letter to Tschirnhaus*, May, 1678, Loemker, 296.
[36]*Letter to Gabriel Wagner*, 1696, Loemker, 763. See further chapter IV, 86, of the present book.
[37]*Reg.*, VIII, H. R., I, 26.
[38]*Ibid.*, IX, H. R., I, 29.

From this theory of cognition there necessarily follows Descartes' definition of method.

Method consists entirely in the order and disposition of the objects towards which our mental vision must be directed if we would find out any truth. We shall comply with it exactly if we reduce involved and obscure propositions step by step to those that are simpler, and then starting with the intuitive apprehension of all those that are absolutely simple, attempt to ascend to the knowledge of all others by precisely similar steps.[39]

The absolute identity of the mind in all its acts of cognition determines the identity of the method of the sciences, whatever their objects. If the mathematical sciences occupy any special rôle in relation to this universal method, it is only because they are the "easiest and clearest" of the sciences, and therefore provide the easiest means of acquiring skill in its exercise.

Condillac, too, derives his conception of universal method from the theory of knowledge. This method is analysis which never varies with its objects. Analysis is a function of language, and a well-constructed science is, therefore, only a well-constructed language. Condillac, like Leibniz, regarded algebra as the perfect model for illustrating this theory of method, for algebra is a language, and he found in it "a striking proof that the progress of the sciences depends uniquely on the progress of language, and that well constructed languages alone can give analysis the degree of simplicity and precision of which it is susceptible."[40] Condillac did not, however, derive this theory of method from considering the nature of algebra, but from an inquiry into the origins of knowledge. The seeking of origins meant, for him, going back to infants and primitive peoples. It was in considering knowledge as it originated with them that he found language to be the indispensable condition of the acquisition of *any* knowledge at all. In its humblest origins language is a *method*. "Every language is an analytic method, and every analytic method is a language."[41] Although languages will be diversified according to their objects, and progress in the sciences occurs when a simpler language is substituted for a more complicated language, nevertheless the function which language can perform for the acquisition of knowledge suffers no diversity at all throughout these changes. This

[39]*Ibid.*, V, H. R., I, 14. See further chapter III, 53 ff. of the present book.
[40]*La Logique* (1780), II, vii, Œuvres, II, 409.
[41]*La Langue des calculs* (1798), Preface, Œuvres, II, 419.

one function is revealed in the primitive origins of human knowledge,
before algebra is considered.[42]

Bacon too lays claim to universality for his method of discovery.

It may also be asked . . . whether I speak of natural philosophy only,
or whether I mean that the other sciences, logic, ethics, and politics,
should be carried on by this method. Now I certainly mean what I have
said to be understood of them all; and as the common logic, which governs
by the syllogism, extends not only to natural but to all sciences; so does
mine also, which proceeds by induction, embrace everything. For I form
a history and tables of discovery for anger, fear, shame, and the like; for
matters political; and again for the mental operations of memory, composition and division, judgment and the rest; not less than for heat and cold,
or light, or vegetation, or the like. But nevertheless since my method of
interpretation, after the history has been prepared and duly arranged,
regards not the working and discourse of the mind only (as the common
logic does) but the nature of things also, I supply the mind with such
rules and guidance that it may in every case apply itself aptly to the nature
of things. And therefore I deliver many and diverse precepts in the doctrine of Interpretation, which in some measure modify the method of
invention according to the quality and condition of the subject of the
inquiry.[43]

Bacon allows for diversities of method for different subject-matters,
but, what is more important, these diversities are still only special
adaptations of one and the same method. The reason that the method
of natural philosophy holds for such apparently different sciences as
logic, ethics, and politics, is that these are not sciences co-ordinate
with natural philosophy but, rather, have their "roots" in it. Natural
philosophy is "the great mother of the sciences."[44]

UNITY OF LANGUAGE

Finally something remains to be said about the interest shown in
these centuries in the possibility of a universal language. The purpose
behind the various proposals for such a language, of which the most
developed were those of Bishop Wilkins and Dalgarno, was rather

[42]See chapter v, 93 f., of the present book.

[43]Novum organum, I, cxxvii, Works, VIII, 159.

[44]In an illuminating article, "Francis Bacon and the Science of Jurisprudence"
(Journal of the History of Ideas, XVIII, 3–26), Paul H. Kocher shows that Bacon
did in fact extend the principles of the Novum organum to jurisprudence in his plans
for the reform of the law. It is worth noting that another great philosopher-jurist of
the century, Leibniz, similarly attempted a reform of the law by means of his universal method. See further, chapter iv, 74, of the present book.

the unity of the international society of scientists through its possession of a common language of communication, than the integration
of the sciences themselves by means of language conceived as a logical
instrument of science. Concern with the logical function of language
as opposed to its rôle in communication was most marked of all in
Hobbes, Leibniz, and Condillac, but of these three it was Leibniz
alone who conceived of a common language for the different sciences
which would unify them in a single system of knowledge. But unity
of language, in Leibniz, is not something distinguishable either from
unity of system or from unity of method.[45]

As propagators of enlightenment the Encyclopaedists, too, were
interested in the relation of language and science. The concern shown
by Diderot for the establishment of community of language for the
sciences was simply an aspect of his concern for bringing knowledge
together for public instruction. If each science were to stay within
the confines of its own language, it would remain isolated from the
rest. To help to break down this isolation Diderot introduced into the
Encyclopaedia, besides the cross-references to things, cross-references
to words. But community of language, however important, did not
mean for him any departure from the vernacular nor the making of a
special universal language for the sciences. If science were to be
popular it must use the language of popular intelligibility. Dominated
by the same, if not an even more fervent, concern for popular enlightenment, and also seeking for the same reason a common basis
for the language of the sciences, Condorcet saw great danger in a
special scientific idiom which could, he said, "divide society into two
unequal classes, the one composed of men who, understanding this
language, would possess the key to all the sciences, the other of men
who, unable to acquire it, would therefore find themselves almost
completely unable to acquire enlightenment."[46] But Condorcet was
nevertheless optimistic about the possibility of creating a universal
language which did not have this disadvantage. It would be like algebra, which he described as "the only really exact and analytical
language yet in existence," containing within it "the principles of a

---

[45]See chapter IV, 81 f., of the present book.
[46]*Sketch for a Historical Picture of the Progress of the Human Mind* (1795), tr.
June Barraclough (New York, 1955), 198.

universal instrument applicable to all combinations of ideas."[47] The existence of such a universal language modelled on algebra would mean that men would not first have to possess the language in order to understand a science, but they would, as in algebra, learn the language in learning the science. It would at once be an instrument of precision and an instrument of discovery, growing with the growth of the sciences, and it would be as effortlessly available to all as the language of algebra. Thus, with one of the last of the French Encyclopaedists in his prescriptions for the progress of the human mind, something of the great ambition of Leibniz was brought to life again.[48]

[47]*Ibid.*, 149.
[48]See H. B. Acton, "The Philosophy of Language in Revolutionary France" (*Proceedings of the British Academy*, XLV, 199–219), for an account of the French controversy over the rôle of language in the acquisition of the sciences. Acton shows how the controversy had its origins in Condillac's theory of the relation of language to ideas.

II. BACON:
NATURAL PHILOSOPHY,
THE GREAT MOTHER
OF THE SCIENCES

THE RECOVERY OF UNIVERSALITY OF KNOWLEDGE

In Bacon's early manuscript, *Valerius Terminus* (1603), there is a section entitled, "Of the impediments of knowledge in handling it by parts, and in slipping off particular sciences from the root and stock of universal knowledge." Under this heading he writes:

Cicero, the orator, willing to magnify his own profession, and thereupon spending many words to maintain that eloquence was not a shop of good words and elegancies but a treasury and receipt of all knowledges, so far forth as may appertain to the handling and moving of the minds and affections of men by speech, maketh great complaint of the school of Socrates; that whereas before his time the same professors of wisdom in Greece did pretend to teach an universal *Sapience* and knowledge both of matter and words, Socrates divorced them and withdrew philosophy and left rhetoric to itself, which by that destitution became but a barren and unnoble science. And in particular sciences we see that if men fall to subdivide their labours, as to be an oculist in physic, or to be perfect in some one title of the law, or the like, they may prove ready and subtile, but not deep or sufficient, no not in that subject which they do particularly attend, because of that consent which it hath with the rest. And it is a matter of common discourse of the chain of sciences how they are linked together, insomuch as the Grecians, who had terms at will, have fitted it of a name of *Circle Learning*. Nevertheless I that hold it for a great impediment towards the advancement and further invention of knowledge, that particular arts and sciences have been disincorporated from general knowledge, do not understand one and the same thing which Cicero's discourse and the note and conceit of the Grecians in their word *Circle Learning* do intend. For I mean not that use which one science hath of another for ornament or help in practice, as the orator hath of knowledge of affections for moving, or as military science may have use of geometry for fortifications; but I mean it directly of that use by way of supply of light and information which the particulars and instances of one science do yield and present for the framing or correcting of the axioms of another science in their very truth and notion. And therefore that example of *oculists* and *title lawyers* doth come nearer my conceit than the other two; for sciences distinguished have a dependence upon universal knowledge to be augmented and rectified by the superior light thereof, as well as the parts and members of a science have upon the *Maxims* of the same science, and the mutual light and consent which one part receiveth of another.[1]

Bacon then gives a number of examples of how one science can throw light on another; natural philosophy on astronomy, astronomy on natural philosophy, natural philosophy on ethics, grammar on logic and rhetoric, music on rhetoric, etc. And in conclusion he remarks that that may be

. . . obscure in the one, which is more apparent in the other, yea and that discovered in the one which is not found at all in the other, and so one science greatly aiding to the invention and augmentation of another. And therefore without this intercourse the axioms of sciences will fall out to be neither full nor true; but will be such opinions as Aristotle in some places doth wisely censure, when he saith *These are the opinions of persons that have respect but to a few things.* So then we see that this note leadeth us to an administration of knowledge in some such order and policy as the king of Spain in regard of his great dominions useth in state; who though he hath particular councils for several countries and affairs, yet hath one council of State or last resort, that receiveth the advertisements and certificates from all the rest.[2]

To this "universal knowledge" Bacon gives the name "*Philosophia Prima*, Primitive or Summary Philosophy" in the *Advancement of Learning*, where he repeats the complaint of the *Valerius Terminus*. "Another error . . . is that after the distribution of particular arts and sciences, men have abandoned universality, or *philosophia prima*; which cannot but cease and stop all progression. For no perfect discovery can be made upon a flat or a level: neither is it possible to discover the more remote and deeper parts of any science, if you stand

[1]*Valerius Terminus*, VIII, *The Works of Francis Bacon*, ed. Ellis, Spedding, and Heath (Boston, 1864), VI, 42–4.
[2]*Ibid.*, 46.

but upon the level of the same science, and ascend not to a higher science."[3]

In the *Advancement of Learning* (1605) and in the *De augmentis* (1623) *philosophia prima* appears in the classification of the sciences as prior to the first division of Philosophy, the division into knowledge of God, knowledge of nature, and knowledge of man. Unlike the others, it is opposed to no other science, "for it differs from the rest rather in the limits within which it ranges than in the subject matter; treating only of the highest stages of things." Bacon compares distributions of knowledge to the branches of a tree which have one stem that grows for some distance before it divides; "therefore it is necessary before we enter into the branches . . . to erect and constitute one universal science, to be the mother of the rest, and to be regarded in the progress of knowledge as portion of the main and common way, before we come where the ways part and divide themselves."[4]

Bacon gives to *philosophia prima* two parts. It is, first, "a receptacle of all such axioms as are not peculiar to any of the particular sciences, but belong to several of them in common."[5] The axioms form a heterogeneous collection, of which the following are some examples: "If equals be added to unequals the wholes will be unequal" is a rule holding in mathematics and in that part of ethics concerned with distributive justice. "Things that are equal to the same thing are equal to one another" holds in mathematics and is the basis of the syllogism. "Things are preserved from destruction by bringing them back to their first principles" is a rule in physics and in politics. "A discord ending immediately in a concord sets off the harmony" is a rule in music which holds in ethics and the affections. Other axioms are cited which hold for music and rhetoric, for perspective and acoustics. These are not to be regarded, says Bacon, as mere "similitudes . . . but plainly the same footsteps of nature treading or printing upon different subjects and matters." He remarks in conclusion that a body of such axioms, which has never yet been collected is "a thing of excellent use for displaying the unity of nature; which is supposed to be the true office of Primitive Philosophy."

It may be observed that Bacon's *philosophia prima* does not consist

[3]*Of the Proficience and Advancement of Learning, Divine and Human,* I, *Works,* VI, 131 f.
[4]*De dignitate et augmentis scientiarum,* III, i, *Works,* VIII, 471.
[5]*Ibid.,* 472.

in principles common to all the sciences, but of principles which are shared by two or more of them. They have therefore a different function from universal principles such as those, for example, in Aristotle's First Philosophy. These latter, of which the most important is the principle of contradiction, do not make possible the transition from one science to another, and, indeed, according to Aristotle, there can be no such transition, for each science marks off some genus and is concerned to demonstrate the essential attributes of that genus, and it is impossible in demonstration to pass from one genus to another. For Bacon, on the other hand, the divisions of the sciences must never be used for separating them from one another, but only as lines to mark them, "in order that solution of continuity in sciences may always be avoided. For the contrary thereof has made particular sciences to become barren, shallow and erroneous."[6] If we may judge from the *Valerius Terminus*, the primary function of the axioms of First Philosophy is to make possible this "continuity" or transition from one science to another so that "the particulars and instances of one science" will supply information for "the framing and correcting of the axioms of another science in their very truth and notion."[7]

Secondly, *philosophia prima* is a doctrine of transcendentals, such as much, little; like, unlike; possible, impossible; being and not being; etc. Here as elsewhere, when Bacon adapts old terms to new purposes, it becomes necessary for him to distinguish his doctrine of transcendentals from the "common" or traditional one. In scholastic terminology the transcendentals were those attributes which transcend the categories of Aristotle and apply to all beings. Six were commonly named: *ens, res, aliquid, unum, bonum, verum*. These have little in common with Bacon's list. There is, however, a classification of transcendentals in Duns Scotus which appears to have some bearing on Bacon's doctrine. For Scotus being itself has the status of being the first of the transcendentals. There then follow the attributes, unity, truth, goodness, which are co-extensive or convertible with being. Thirdly, there are the disjunctive attributes: infinite–finite, substance–accident, necessary–contingent, actual–potential, etc.— Scotus presents no exhaustive list of them. These attributes, when taken in disjunction, are coextensive with being. Thus, for example,

[6]*Ibid*., IV, i, *Works*, IX, 14.
[7]*Val. Term*., VIII, *Works*, VI, 44.

every being is actual or potential, necessary or contingent, substance or accident, etc.[8] Bacon's "Much, Little; Like, Unlike; Possible, Impossible; likewise Being and Not Being, and the like" plainly belong in the class of disjunctive transcendentals. There are presumably for him no convertible transcendentals, and being, which for Scotus was the first of the transcendentals, is reduced to the status of membership in a disjunctive pair. Bacon's first philosophy is therefore not a metaphysics of being *qua* being.

It is also relevant to note that for the scholastics logic and metaphysics both had their transcendentals, in accordance with the distinctions which they commonly made between real and logical concepts, or between first and second intentions. The former concepts are predicable directly of things, the latter are predicable only of other concepts. Bacon takes account of these two ways of treating transcendentals when he remarks on "the distinction which is current, that the same things are handled but in several respects; as for example, that logic considereth of many things as they are in notion, and this philosophy [*philosophia prima*] as they are in nature: the one in appearance, the other in existence. But I find this difference better made than pursued."[9] If, however, men had considered the transcendentals "as philosophers, and in nature, their inquiries must of force have been of a far other kind than they are. . . . But there is a mere and deep silence touching the nature and operation of those Common Adjuncts of things, as in nature; and only a resuming and repeating of the force and use of them in speech or argument."[10] Transcendentals must be "handled as they have efficacy in nature and not logically."[11] Thus, for example, in the case of much and little, one might inquire why some things in nature are so plentiful and others so scarce; or in the case of like and unlike, why there are always some individuals in nature which fall between two species, partaking of the nature of both, or why gold does not attract gold, which is like it, but does attract quicksilver which is unlike it.

There is another misconception latent in the use of old terminology from which *philosophia prima* must be rescued. Not only is it not, as we have seen, a metaphysics of being *qua* being; it is not a meta-

---

[8]See A. B. Wolter, *The Transcendentals and their Function in the Metaphysics of Duns Scotus* (Washington, 1946).
[9]*Adv.*, II, *Works*, VI, 209.
[10]*Ibid.*
[11]*De aug.*, III, iv, *Works*, VIII, 484.

physics of any kind, although traditionally the terms "first philosophy" and "metaphysics" have been taken as equivalent. Metaphysics, for Bacon, is a part of natural philosophy, coming in the order of induction after physics, whereas *philosophia prima* is prior to all the sciences, including natural philosophy.

Between the *Advancement of Learning* (1605), and the translation and revision of it which appeared in the *De augmentis* (1623), there appeared the major work, the *Novum organum* (1620), a product of Bacon's mature thought. This work makes no mention of *philosophia prima*. There are, nevertheless, passages in it which possess a direct connection with what had been said on this science in the *Advancement of Learning* and which was to be repeated in the *De augmentis*. In referring in the *Novum organum* once more to the causes of the "errors" of philosophers, Bacon includes among them the fact that "during those very ages in which the wits and learning of men have flourished most, or indeed flourished at all, the least part of their diligence was given to natural philosophy. Yet this very philosophy it is that ought to be esteemed the great mother of the sciences. For all arts and all sciences, if torn from this root, though they may be polished and shaped and made fit for use, yet they will hardly grow."[12] In the next aphorism, referring once again to "this great mother of the sciences," he says,

Meanwhile let no man look for much progress in the sciences—especially in the practical part of them—unless natural philosophy be carried on and applied to particular sciences, and particular sciences be carried back again to natural philosophy. For want of this, astronomy, optics, music, a number of mechanical arts, medicine, itself—nay, what one might more wonder at, moral and political philosophy, and the logical sciences, —altogether lack profoundness and merely glide along the surface and variety of things; because after these sciences have been once distributed and established, they are no more nourished by natural philosophy.—And therefore it is nothing strange if the sciences grow not, seeing they are parted from their roots.[13]

In a still later aphorism he reverts once more to this point. "And here also should be remembered what was said above concerning the extending of the range of natural philosophy to take in the particular sciences, and the referring or bringing back of the particular sciences to natural philosophy, that the branches of knowledge may not be

[12]*Novum organum*, I, lxxix, *Works*, VIII, 110.
[13]*Ibid.*, I, lxxx, *Works*, VIII, 112.

severed and cut off from the stem. For without this the hope of progress will not be good."[14]

In these passages there is an exact repetition of some of the expressions previously referred in the *Advancement of Learning* to *philosophia prima* but now referred to natural philosophy. It is now natural philosophy which is described as the "mother of the sciences," and as "the stem" of which the other sciences are the "branches." Where in the *Valerius Terminus* Bacon had referred to "slipping off particular sciences from the root and stock of universal knowledge," he now describes the separating of the particular sciences from natural philosophy as the parting of them "from their roots." Moreover, the same claim is made for natural philosophy which previously was made for *philosophia prima*, that if the particular sciences are cut off from it all progress in them will cease. Lastly, it may be noted that where formerly it was *philosophia prima* which was the general science distinguished from the *particular* sciences, of which natural philosophy was only one, it is now in the *Novum organum* natural philosophy which is the general science distinguished from the particular sciences. There is no reason to say that Bacon had undergone a change in his views as to what unites all the particular sciences, for the *De augmentis* which appeared after the *Novum organum* reiterates the earlier theory concerning *philosophia prima* and even expands it, though not altering any of its essential features. In spite, however, of these differences in the two accounts, there is a basic consistency underlying the theory of *philosophia prima* and the forthright naturalism of the *Novum organum*. Bacon's *philosophia prima* belongs within a philosophy in which all knowledge is restricted to nature as its object. The axioms of *philosophia prima* are, he has said, "plainly the same footsteps of nature treading or printing upon different subjects and matters." "The true office of Primitive Philosophy" is that of "displaying the unity of nature." And again, what he claims of novelty for his mode of treating trancendentals is that they are "handled as they have efficacy in nature." Although it is true that in the *Advancement of Learning* he distinguishes divine, natural, and human philosophy, yet in the same work he acknowledges that the knowledge of man, to the extent to which he can be subject to scientific inquiry, "is but a portion of natural philosophy in the

[14]*Ibid.*, I, cvii, *Works*, VIII, 139 f.

continent of nature."[15] As for natural theology or divine philosophy, it can, according to Bacon, impart no information about the nature of God, but is confined to his observable works in nature. In any case, whether we take the account given in the *Advancement of Learning* or in the *Novum organum*, each indicates the same thing, namely that the basis for Bacon's conception of the unity of the sciences is a thorough-going philosophical materialism, for nature consists of a wholly self-determinate matter, and to a consideration of this materialism and its consequences for the sciences we must now turn.

NATURAL PHILOSOPHY AS IMAGING THE SELF-SUFFICIENCY AND UNITY OF NATURE

Matter is the principal subject in Bacon's interpretation of the ancient fable of Cupid, as presented in his *De principiis atque originibus* (1623–4) and *De sapientia veterum* (1609). The fable shows that considered in itself matter can have no cause. It belongs to no genus. "Wherefore," says Bacon, "whatsoever this matter and its power and operation be, it is a thing positive and inexplicable, and must be taken absolutely as it is found, and not to be judged by any previous conception."[16] It cannot be known by a cause, for as the cause of all natural phenomena, it is itself without a cause. Scientific inquiry cannot go beyond it, and though revelation informs us that God created the world, God himself lies completely outside nature and there can be no argument within the chain of causes to God as First Cause.

Within the confines of nature itself, it is hardly to be hoped for that man can attain knowledge of the ultimate method by which matter operates. "For the summary law of nature, that impulse of desire [or force[17]] impressed by God upon the primary particles of matter which make them come together, and which by repetition and multiplication produces all the variety of nature, is a thing which mortal thought may glance at, but can hardly take in."[18] This summary law represents the absolute terminus to which natural inquiry can theoretically, if not practically, attain. By no steps can we ascend

[15]*Adv.*, II, *Works*, VI, 236.
[16]*De principiis atque originibus*, *Works*, X, 345.
[17]*Ibid.*
[18]*De sapientia veterum*, XVII, *Works*, XIII, 123.

from nature to the divine. Those who have referred the original impulse of matter—or "natural motion of the atom"—to God "ascend by a leap and not by steps."[19] Of any relation which God has to nature we can know only through Sacred Writ. It cannot be learned from nature itself. First, that matter was created by God out of nothing is something which we know by faith alone; "for by one who philosophizes according to sense alone, the eternity of matter is asserted." Secondly, that the development of the orderly system of the universe out of the original chaos of fully formed matter was by Divine Omnipotence, and was not the result of the blind operation of matter itself, is similarly not known from natural inquiry. Nor is it possible to know that the order of the universe existing before the Fall was the best to which matter is susceptible. "In these points therefore we must rest upon faith and the firmaments of faith."[20]

Bacon does not deny that God acts providentially in nature. But he does deny that God imposes any of the character of his providence upon the workings of nature. There is not even any coincidence between the activities of nature and the providential activities of God in nature—rather they are contrary.

For as in civil actions he is far the greater and deeper politician that can make other men the instruments of his ends and desires and yet never acquaint them with his purpose (so that they shall do what he wills and yet not know that they are doing it) than he that imparts his meaning to those he employs; so does the wisdom of God shine forth more admirably when nature intends one thing and Providence draws forth another, than if he had communicated to all natural figures and motions the characters and impressions of his providence.[21]

Thus nothing of the divine will exists in nature nor is imparted to it. God is as distinct from nature as the workman from his work. If the evidence of design in nature must lead to the acknowledgment of God's existence, it is precisely because design is not the work of nature herself—"Aristotle, when he had made nature pregnant with final causes . . . had no further need of God."[22]

Just as an artifact can reveal the power and skill of the workman, but does not reveal his image, so also with the works of God. Nature

19Ibid.
20De princ., Works, X, 386.
21De aug., III, iv, Works, VIII, 511.
22Ibid.

cannot reveal the likeness of God. "There is no proceeding in invention of knowledge but by similitude; and God is only self-like, having nothing in common with any creature."[23] The world, then, is not as certain heathen philosophers have thought, the image of God, and the Scriptures have never said so, saying only that it is the work of His hands.[24]

This absolute separation of the divine from the natural means that philosophy or science can have no object but nature or matter. It means also that all scientific knowledge will be derived from the senses, for it is only through the senses that nature acts on the spirit. "Sense . . . is the reflection of things material." "God has framed the mind of man as a glass capable of the image of the universal world, joying to receive the signature thereof as the eye is of light."[25] But this world of nature is all that the senses are capable of imaging. For knowledge of the divine we are dependent wholly on revelation. We must, therefore, at all times distinguish between "the light of nature" and "the light of divinity,"[26] between "the oracles of sense" and "the oracles of faith."[27] There must be no intrusion of natural philosophy into sacred theology, and with equal stringency it is asserted that there must be no intrusion of sacred theology into philosophy.[28] To mix them is "to have at once a heretical religion and a fabulous philosophy."[29]

The unity of nature is a principal subject in Bacon's interpretation of the fable "Pan, or Nature." Pan represents "the universe, or the all of things." The fable attributes no loves to Pan except his marriage with Echo. This absence of love for anything but an insubstantial thing is interpreted as signifying the perfect self-sufficiency and completeness of nature. "For the world enjoys itself, and in itself all things that are." The world lacks nothing, it desires nothing, and it has no issue. "Generation goes on among the parts of the world, but how can the whole generate, when no body exists out of itself?"[30]

Within this one self-sufficient world of nature all its variety is produced by a single principle, that impulse of desire or force original

[23]Val. Term., I, Works, VI, 29.
[24]De aug., III, ii, Works, VIII, 478.
[25]Val. Term., I, Works, VI, 32.
[26]De sap., X, Works, XIII, 109.
[27]Ibid., XXVI, Works, XIII, 155.
[28]Nov. org., I, lxv, Works, VIII, 93 f.
[29]De sap., XXVI, Works, XIII, 155.
[30]Ibid., VI, Works, XIII, 101.

in the primary particles of matter, which is expressed in the summary law of nature. This aspect of nature's unity is revealed in the part of the fable which refers to Pan's horns. "Horns are attributed to the universe, broad at the base and pointed at the top. For all nature rises to a point like a pyramid. Individuals, which lie at the base of nature, are infinite in number; these are collected into Species, which are themselves manifold; the Species rise again into Genera; which also by continual gradations are contracted into more universal generalities, so that at last nature seems to end as it were in unity; as is signified by the pyramidal form of the horns of Pan."[31]

It was noted previously that the only love of Pan was his marriage with Echo—"a thing not substantial but only a voice." "But it is well devised that of all words and voices Echo alone should be chosen for the world's wife; for that is the true philosophy which echoes most faithfully the voices of the world itself, and is written as it were at the world's own dictation; being nothing else than the image and reflexion thereof to which it adds nothing of its own, but only iterates and gives it back."[32]

Since philosophy is merely the imaging of nature, it will exhibit nature's unity. The sciences of nature will be so ordered in relation to one another as to reflect the ascent from the multiplicity of individual things to their ultimate unity in the summary law of nature. That is to say, these sciences too will form a pyramid. At the base of the pyramid is natural history. On that is built physics, which has two parts, one less general ("Physic concerning things concrete") and one more general ("Physic concerning things abstract"). On physics is built metaphysics which subsumes the axioms of physics under still more general axioms. "As for the cone and vertical point (. . . namely, the summary law of nature) it may fairly be doubted whether man's inquiry can attain to it." The three inductively ordered levels, natural history, physics, and metaphysics are "the true stages of knowledge" and Bacon finds "the speculation was excellent in Parmenides and Plato (although in them it was but a bare speculation) 'that all things by a certain scale ascend to unity'. So then always that knowledge is worthiest which least burdens the intellect with multiplicity; and this appears to be metaphysic. . . ."[33] The simple

[31]De aug., II, xiii, Works, VIII, 449.
[32]Ibid., 456.
[33]Ibid., III, iv, Works, VIII, 507.

forms of things, which are its objects, constitute an alphabet of nature. Like the letters of other alphabets they are limited in number. The various ways in which they are combined account for everything in nature and make all its variety. But a second respect which ennobles metaphysics consists in the extensive power and freedom which it bestows upon man for the control of nature, a power far greater than that made possible by physics. Where physics produces mechanics, metaphysics produces magic, not "the popular and degenerate natural magic," but "the science which applies the knowledge of hidden forms to the production of wonderful operations."[34]

The ascent to unity belongs entirely within the realm of matter. Neither the use of the term "metaphysic" nor the reference to Platonic doctrine should be allowed to obscure Bacon's conception of it. Bacon praises Plato for having taught that Forms are the true objects of knowledge but, however right Plato was, he erred in seeking to apprehend Forms in absolute abstraction from matter, "whence it came," says Bacon, "that he turned aside to theological speculations, wherewith all his natural philosophy is infected and polluted."[35] Forms in abstraction are mere fictions of the mind. In nature nothing exists but individual bodies, performing individual acts according to fixed laws. Bacon explains that it is to such laws that he refers when he speaks of "Forms," adapting, as is his custom, old terms to new uses. To know these Forms or laws of motion is to embrace the unity of nature, and it is this knowledge which is metaphysics. Lying beyond these unifying laws, there is, however, a single physical law containing them all, and expressing the ultimate unity of nature. This is the summary law of nature. If, as Bacon says, "Metaphysics . . . is itself a part of Physics, or of the doctrine concerning nature,"[36] no less so is this vertical point of the pyramid of knowledge. There is no ascent through physics to knowledge of the divine. Nature contains all its principles of explanation within itself.

MAN IN RELATION TO NATURE, AND THE SCIENCES OF MAN

The rigid dualism of the divine and the natural or material is extended by Bacon into human nature itself. The division occurs,

[34]*Ibid.*, III, v, *Works*, VIII, 513 f.
[35]*Ibid.*, VIII, 505.
[36]*Ibid.*, IV, iii, *Works*, IX, 59.

however, not between man's soul and his body, but within the soul. Man has a rational soul which is divine, and an irrational soul in common with the brutes. The latter is purely material. This "sensible or produced soul," as Bacon calls it, must be regarded as a "corporeal substance, attenuated and made invisible by heat," a breath or spirit made up of the natures of flame and air.[37] It alone is an appropriate object of philosophic or scientific inquiry. The rational soul itself must be excluded. Man is made in the image of God, as the Scriptures directly say, but as the nature of God remains ultimately mysterious, and hidden, so also does the nature of man, in so far as he possesses a rational soul. All questions concerning this soul must therefore be handed over to religion. Such questions traditionally asked by philosophers, "whether it be native or adventive, separable or inseparable, mortal or immortal, how far it is tied to the laws of matter, how far exempted from them, and the like"[38]—all are removed from philosophy. "For since the substance of the soul in its creation was not extracted or produced out of the mass of heaven and earth, but was immediately inspired from God; and since the laws of heaven and earth are the proper subjects of philosophy; how can we expect to obtain from philosophy the knowledge of the substance of the rational soul? It must be drawn from the same divine inspiration, from which that substance first proceeded."[39]

As existing in nature, but possessing a rational soul which is divine, man occupies a status in nature which is unique. He is "the special and peculiar work of Providence," and indeed is the clue to all the other provident actions of God in nature. "Man, if we look to final causes, may be regarded as the centre of the world; insomuch that if man were taken away from the world, the rest would seem to be all astray, without aim or purpose, to be like a besom without a binding, as the saying is, and to be leading to nothing. For the whole world works together in the service of man; and there is nothing from which he does not derive use and fruit."[40]

It is this supreme status of man in nature which determines for Bacon what the end of scientific knowledge is. We have seen that although God is completely other than nature, he nevertheless acts

[37]*Ibid.*, IX, 50.
[38]*Ibid.*
[39]*Ibid.*
[40]*De sap.*, XXVI, *Works*, XIII, 147.

providentially in nature. The chief of his provident works is man himself, whom he has created in his own image. In endowing man with mind and intellect, God has created a provident being—one whose relation to nature is an image of his own relation to it. When he attributes the first place among human actions to the introduction of inventions and works, Bacon says, "discoveries are, as it were, new creations, and imitations of God's works."[41] As supreme among the creatures in nature, man is endowed with "a right over nature," and Nature is variously described by Bacon as "the kingdom of man" or "the empire of man." Nevertheless, though he is given the right to this kingdom, it is necessary for man himself to assert and establish his dominion over the universe. In the first stage of his existence he is "a naked and defenceless thing, slow to help himself and full of want." He is not endowed with the actual control of nature, but with the capacity and right to attain that control through his own efforts. It is necessary that he be continually dissatisfied with the knowledge that he already possesses. Perpetual complaint of the existing arts and sciences is the stimulus to "fresh industry and new discoveries." "Conceit of plenty is one of the principal causes of want."[42] The work of science is based on ceaseless dissatisfaction and striving and is essentially concerned with "the discovery of particulars not revealed before for the better endowment and help of man's life."[43]

Just as it is religion which sets the limits to scientific knowledge, so also it determines the end of such knowledge. For if God has given man the whole world of nature for inquiry, "yet evermore it must be remembered that the least part of knowledge passed to man by this so large a charter from God must be subject to that use for which God hath granted it; which is the benefit and relief of the state and society of man." Scientific knowledge has no merit except as referred to Christian charity. St. Paul "doth notably disavow both power and knowledge such as is not dedicated to goodness or love, for saith he, *If I have all faith so as I could remove mountains* (there is power active), *if I render my body to the fire* (there is power passive), *if I speak with the tongues of men and angels* (there is knowledge, for language is but the conveyance of knowledge), *all were nothing.*"[44]

[41]*Nov. org.*, I, cxxix, *Works*, VIII, 161.
[42]*De sap.*, XXVI, *Works*, XIII, 150.
[43]*Val. Term.*, IX, *Works*, VI, 50.
[44]*Ibid.*, I, *Works*, VI, 34.

The obligation of charity prohibits a "mere contemplation which should be finished in itself without casting beams of heat and light upon society."[45]

What now are the effects of this on the sciences specifically concerned with man himself? Bacon acknowledges that man is of all things that which concerns us most, but of nature man is but a part.[46] The study of man must be subordinated to the study of nature as a whole, and Bacon was deeply critical of those philosophers who had turned away from nature to preoccupy themselves with ethics and politics. Their treatment of human nature in abstraction from natural philosophy could only be superficial.

But Bacon has already placed a severe restriction upon the sciences of man. In man's nature, that part of him which is divine, or made in the image of God, lies beyond the competence of scientific inquiry. What then remains as possible for man's philosophic knowledge of himself? Eliminating from the sciences of man those specifically concerned with his body in Bacon's classification, such as medicine, athletic, cosmetic, and the voluptuary arts, there remain three principal ones to be considered—logic, ethics, and politics, and two of these, logic and ethics, are designated by Bacon as sciences of man's mind. Though the substance of the rational soul has been removed from inquiry, nevertheless the faculties of the rational soul have their employment in nature. The "use and objects" of these faculties are therefore legitimate objects of inquiry. "Logic discourses of the Understanding and Reason; Ethic of the Will, Appetite and Affections: the one produces determinations, the other actions."[47]

LOGIC

The nature and value of the new logic of induction derives directly from the goal which Bacon has set before scientific knowledge. It is only this logic which can have as its effect "to command nature in action." "I consider induction," he says, "to be that form of demonstration which upholds the sense and closes with nature, and comes to the very brink of operation, if it does not actually deal with it."[48] Bacon's logic makes no incursion into the nature of the rational soul,

[45]De aug., VII, i, Works, IX, 199.
[46]Ibid., IV, i, Works, IX, 14.
[47]Ibid., V, i, Works, IX, 61.
[48]The Great Instauration, Plan of the Work, Works, VIII, 42.

and rests on no theory about it, as does, for example, Descartes'. It is limited to the observable use of the faculties of understanding and reason, and is concerned only to direct and strengthen that use. "It is the duty of Art to perfect and exalt Nature. . . . For he that shall attentively observe how the mind doth gather this excellent dew of knowledge . . . distilling it and contriving it out of particulars natural and artificial . . . shall find that the mind of herself by nature doth manage and act on induction much better than they [the 'logicians'] describe it."[49] But at the same time the mind also suffers from certain natural impediments, and it is the purpose of the new logic to free the mind from these. "There remains but one course for the recovery of a sound and healthy condition, namely that the entire work of the understanding be commenced afresh, and the mind itself be from the very outset not left to take its own course, but guided at every step; and the business be done as if by machinery."[50] The logic of discovery is a technical device applied from without to the mind. It has exactly the same purely instrumental function of control in relation to the understanding which, as we shall see, his "Georgics of the Mind" have in relation to the will in ethics, and which natural philosophy has in relation to nature.

ETHICS

In defending the cause of learning Bacon had to take account of the Biblical doctrine that the original temptation and sin by which man fell was the aspiration to too much knowledge. It was not, he pointed out, the aspiring to knowledge of nature which was the occasion for the fall. No part of the natural world is denied to man's inquiry, and there is no limitation upon the extent to which man may extend his physical researches. "It was the ambitious and proud desire of moral knowledge to judge of good and evil, to the end that man may give laws to himself, which was the form and manner of the temptation."[51] The supposition behind this desire was that good and evil did not have their origin in the commands of God, but had other foundations, and that if these could be discovered man could depend wholly upon himself.

The moral law belongs, along with the great mysteries which

[49]*Adv.*, II, *Works*, VI, 265.
[50]*Nov. org.*, Preface, *Works*, VIII, 61.
[51]*The Great Instauration*, Preface, *Works*, VIII, 35 f.

concern the Deity, to Sacred Theology. The greater part of it "is higher than the light of nature can aspire to." It has its sources not in "the dictates of reason," but in divine will as revealed in God's word. Bacon grants that Christian revelation is not the only source of knowledge of virtue and vice. There is a kind of natural revelation, distinct from sense and induction, an inward instinct or conscience, by which the soul is able to some extent to perceive the moral law. But this moral knowledge lies outside the sphere of philosophy. It is not, Bacon points out, got from "sense, induction, reason, argument, according to the laws of heaven and earth."[52] It is a relic of man's original purity.

If, then, ethics has any place among the sciences, it can have only a severely limited function. It "may be admitted into the train of theology, as a wise servant and faithful handmaid to be ready at her beck to minister to her service and requirements." Moral philosophy must always subordinate itself to the doctrines of divinity, but it may within its own limits yield "many sound and profitable directions."[53]

Ethics has been defined by Bacon as the doctrine concerning the use and objects of the faculties of will, appetite, and affections. Like logic it is essentially a science for the control and directing of the appropriate faculties. Bacon finds that an ethics, scientific in the sense in which he conceives it, has been almost wholly neglected by the moralists of the past.

In the handling of this science, those which have written seem to me to have done as if a man that professed to teach to write, did only exhibit fair copies of alphabets and letters joined, without giving any precepts or directions for the carriage of the hand and framing of the letters. So have they made good and fair exemplars and copies, carrying the draughts and portraitures of *good, virtue, duty, felicity*; propounding them well described as the true objects of man's will and desires. But how to attain these excellent marks, and how to frame or subdue the will of man to become true and conformable to these pursuits, they pass it over altogether, or slightly and unprofitably.[54]

Ethics is given two parts by Bacon. One is concerned with the nature of the good, the other "prescribing rules how to accommodate the will of man thereto." The former of these without the latter "seems to be

52*De aug.*, IX, i, *Works*, IX, 348.
53*Ibid.*, VII, iii, *Works*, IX, 215.
54*Adv.*, II, *Works*, VI, 309.

no better than a fair image or statue, which is beautiful to contemplate, but is without life and motion. . . ."[55] As to questions concerning the nature of the *summum bonum*, over which the heathen philosophers infinitely disputed and speculated, they "are by the Christian faith removed and discharged." Within the science of ethics, therefore, it is the Georgics of the mind which assumes importance. Their task is "really to instruct and suborn action and active life."[56]

This culture of the mind for subduing, applying, and accommodating the will of man to the good rests on a knowledge of "characters of disposition" and a science of the "affections" or passions. It is history which supplies the material for such knowledge. All knowledge exists for the sake of action, and it is this end which gives to history its significance, to civil history no less than to natural history. Bacon never tires of saying that in order to command nature we must obey her and by obeying nature he means learning from her by observation. "For the matter in hand is no mere felicity of speculation, but the real business and fortunes of the human race, and all power of operation. For man is but the servant and interpreter of nature; what he does and what he knows is only what he has observed of nature's order in fact or in thought; beyond this he knows nothing and can do nothing."[57]

POLITICS

The nature of the restriction imposed upon the scope of ethics does not extend to politics, for the two sciences are completely separated from one another by Bacon. This divorce means in the first place that politics does not concern itself with questions about the nature and grounds of the obligations of the subject to the state, nor with questions about a prince's obligations and the source of the legitimacy of his authority. These are questions which Bacon firmly classifies as ethical, not political, and, as ethical, they are subject to all the divinely determined limitations within which a science of ethics is permitted to proceed. Ethics, in another of his divisions of that science, contains two parts, one of which is concerned with "Individual or Self-Good," the other with the "Good of Communion." The latter, says Bacon, "may seem at first glance to pertain to science civil and politic, but

[55]*De aug.*, VII, iii, *Works*, IX, 214.
[56]*Adv.*, II, *Works*, VI, 311.
[57]*The Great Instauration*, Plan of the Work, *Works*, VIII, 53.

not if it be well observed; for it concerns the regimen and government of every man over himself, and not over others."[58] Included within this part of ethics are " 'the common duty of every man' as a member of the state," and the special duties which belong to the exercise of authority. It is within this part of ethics, not under politics, that Bacon classifies King James's two works, the *Basilicon Doron*, which discourses on a king's Christian duty towards God and his duty to his office, and *The True Law of Free Monarchies*, a formulation of the doctrine of the Divine Right of Kings, in which, as Bacon observes to James, "it well appears that you no less perceive and understand the plenitude of the power of a king and the ultimities . . . of legal rights, than the circle and bounds of his office and duty."[59]

Besides, however, the removal of these matters from politics, Bacon's separation of the two sciences produces another consequence of radical import. Aristotle supposed that if man were "the best thing in the world"—and this he denied—politics would be the best knowledge. It is central to Bacon's whole treatment of the sciences that man is the best thing in the world, but unlike Hobbes he signally fails to verify the prediction implied in Aristotle's statement. On the contrary, he explicitly asserts the superiority of the mechanic arts to the political arts. This evaluation of their relative merits rests on Bacon's conception of the state as a phenomenon, possessing certain natural appetites and desires in common with the rest of nature. The state is not related to the life of man in such a way as to give it an authoritative rôle in the ordering of all human pursuits. As natural it fails to encompass that aspect of man's nature which is outside the natural order and which belongs to the divine. The "good for man" does not fall under the surveillance of philosophy, but the "good for the state" does, since this good is determined solely by the state's natural appetites. The grounds on which politics traditionally had claimed an architectonic status among the practical sciences do not exist for Bacon.

It is worth while to recall Aristotle's classic argument in order to place Bacon's position in full light. Every art, says Aristotle, is thought to aim at some good. But there are many arts, and there are therefore many ends. These varieties of ends are not, however, autonomous, but are hierarchically ordered in relation to one another. Just as bridle-

[58]*De aug.*, VII, ii, *Works*, IX, 207.
[59]*Ibid.*, 210.

making and the other arts concerned with the equipment of horses fall under the art of riding, so other arts fall under yet others. It is for the sake of the latter that the former are pursued. But if, Aristotle continues, there is something which is good in itself, this would be the chief good, and it would be the subject of "the most authoritative art and that which is most truly the master art. And politics appears to be of this nature . . . now, since politics uses the rest of the sciences, and since, again, it legislates as to what we are to do and what we are to abstain from, the end of this science must include those of the others, so that this end must be the good for man."[60]

The end of the state is for Aristotle the good life and this conception was never seriously challenged until the Renaissance. Throughout the Middle Ages the state continued in philosophic theory to be directed towards the virtuous life of its citizens. The challenge came from the politics of reason of state. The fundamental assumption as reflected, for example, in the writings of Machiavelli, is that the state's ends are autonomous; they are its own and distinct from moral ends. Politics is the technique for attaining these ends and it is a privilege of princes, for it is concerned with the power of princes, with the maintenance of that power and with its extension. If the state seeks the security and prosperity of its subjects, it is not because justice and happiness present themselves as moral requirements, but because the strength of the state rests on the security and prosperity of its subjects. And if the state must seek to expand its power it is because it is in the nature of states to do so.

In separating politics completely from ethics Bacon avowedly breaks with the tradition of classical and mediaeval philosophy. "Civil knowledge, which is commonly ranked as a part of ethic I have . . . emancipated and erected into an entire doctrine by itself."[61] Politics becomes identified now with the arts or techniques of exercising power. The ends of these arts consist in the satisfaction of certain appetites which are natural to the state as such. "There is impressed on all things a triple desire or appetite, in respect of self or individual good; one of preserving, another of perfecting, and a third of multiplying and spreading themselves."[62] This triple desire or appetite determines the ends of "wisdom of state." "The Arts of Government contain three

[60]*Ethics*, I, ii, tr. W. D. Ross (Oxford, 1942).
[61]*De aug.*, V, i, *Works*, IX, 60 f.
[62]*Ibid.*, VII, ii, *Works*, IX, 204.

political duties; first 'the preservation,' secondly, 'the happiness and prosperity,' and thirdly, 'the extension' of empire."[63] The last of these appetites, that for aggrandizement, is the worthiest, because the most active. In the extension of empire lies the true greatness of kingdoms and states.

To sever, as Bacon does, politics from ethics, is to demolish the basis for the supremacy of politics among the arts. It remains an independent art, one among others, but like them grounded on natural philosophy, which discovers the same appetites and desires manifested in all things.

In conclusion it would be well to reiterate the interconnected rôles of philosophy and religion in determining the unity of the sciences for Bacon. As a philosopher Bacon conceives of nature as wholly self-sufficient, requiring no principle of explanation outside itself. Consequently natural philosophy is discontinuous with any other kind of inquiry. But as to what other kinds of scientific inquiry are possible, Bacon is without either a metaphysics or an epistemology which can pronounce on such a question. It is religion and scriptural injunction which perform this function. Since his metaphysics is merely knowledge of the most general laws governing the motions of natural bodies, it can say nothing about kinds of knowledge. As for his epistemology, it is, as Professor Anderson has pointed out, almost non-existent.[64] In giving to the opening section of his *Valerius Terminus* the title "Of the limits and end of knowledge" Bacon is not, as might appear, posing a familiar philosophical problem, for under this heading he writes, "I thought it good in the first place to make a strong and sound head or bank to rule and guide the course of the waters; by setting down this position or firmament, namely, *That all knowledge is to be limited by religion, and to be referred to use and action.*"[65] It is religion which legislates that science cannot presume to inquire into the being or attributes of God, into the nature of man's rational soul, or into the basis of moral law and the nature of man's highest good. It is these limits imposed by religion

[63]*Ibid.*, VIII, iii, *Works*, IX, 298 f.

[64]"Bacon's epistemology—as distinct from the rules of induction—all but disappears before the triple demands of a revealed theology, a materialistic psychology, and a mechanical logic." F. H. Anderson, *Philosophy of Francis Bacon* (Chicago, 1948), 214.

[65]*Val. Term*, I, *Works*, VI, 28.

which confine science to the self-sufficient realm of matter and which determine consequently that all legitimate sciences must be branches of natural science. It is also religion which legislates the common end of all sciences—that they are "to be referred to use and action," for it is Scripture, through the voice of St. Paul, which imposes the Christian obligation of charity and requires that knowledge shall be pursued for "the benefit of man and relief of the state and society of man."[66] And if, finally, all legitimate sciences are subject to the one method, this is a consequence of their all being natural sciences and directed to use and action.

[66]Ibid., 38.

III. DESCARTES:
THE PROJECT
OF A UNIVERSAL SCIENCE

UNIVERSAL WISDOM

There are two main statements in Descartes' writings on the unity of the sciences, one in the opening section of his earliest philosophical work, the *Regulae ad directionem ingenii* (1628), and the other in the Preface to the *Principles of Philosophy* (1644). The former of these is introductory to the great project of inquiry on which he was just embarking, setting the goal for it; the latter accompanies a comprehensive presentation of the main results of that inquiry. Both are on the nature of wisdom, with the pursuit of which he identified philosophy.

The first rule in the *Regulae* states: "The end of study should be to direct the mind towards the enunciation of sound and correct judgments on all matters that come before it." This is followed by an attack on the isolating of the various sciences for specialized inquiry. Descartes attributes the existence of such specialization to the influence of a false analogy between science and the manual arts. The latter, because they involve the use of the body, are compelled to be specialized. The ways, for example, in which the hand is adapted to agriculture and to playing the harp are quite distinct, and skill in the former is no aid to the acquisition of skill in the latter but is actually detrimental to it. This well-recognized fact has led men to suppose that "the cognitive exercise of the mind" must also vary according to its different subject-matters, and that, therefore, like the arts, the sciences should be pursued in complete separation from one another.

But this is certainly wrong. For since the sciences taken all together are identical with human wisdom, which always remains one and the same, however applied to different subjects, and suffers no more differentiation proceeding from them than the light of the sun experiences from the variety of the things which it illumines, there is no need for minds to be confined at all within limits; for neither does the knowing of one truth have an effect like that of the acquisition of one art and prevent us from finding out another, it rather aids us to do so.[1]

Descartes finds it strange that people have investigated all sorts of special subjects without ever concerning themselves with "good understanding or universal Wisdom," although ultimately their value consists in what they contribute to this general end. Even the pursuit of special sciences for such praiseworthy motives as the human conveniences they make possible, or the pleasure which is afforded by the sheer contemplation of truth, "practically the only joy in life that is complete and untroubled with any pain," can cause us to overlook certain facts as being of little interest or value, although they may be necessary for the understanding of other things.

Hence we must believe that all the sciences are so inter-connected, that it is much easier to study them all together than to isolate one from all the others. If, therefore, anyone wishes to search out the truth of things in serious earnest, he ought not to select one special science; for all the sciences are conjoined with each other and interdependent: he ought rather to think how to increase the natural light of reason, not for the purpose of resolving this or that difficulty of scholastic type, but in order that his understanding may light his will to its proper choice in all the contingencies of life. In a short time he will see with amazement that he has made much more progress than those who are eager about particular ends, and that he has not only obtained all that they desire, but even higher results than fall within his expectation.[2]

We see, then, that in the *Regulae* Descartes identifies "universal Wisdom" with "good understanding (*bona mens*)" or—as it is described in the formal statement of the rule—with the capacity of the mind to form sound and true judgments on all matters that come before it. It is also given the name "natural light of reason," reference being made once more to the universality of its application. The use of this key expression, *bona mens*, invites comparison with his later

[1]*Regulae ad directionem ingenii*, I, *The Philosophical Works of Descartes*, tr. E. S. Haldane and G. R. T. Ross (Cambridge, I, 1931; II, 1934), I, 1 f.
[2]*Ibid.*

statement on "good sense," with which the *Discourse on Method* (1637) opens, for the Latin translation of the *Discourse* rendered *bon sens* as *bona mens*.[3]

Good sense is of all things in the world the most equally distributed, for everybody thinks himself so abundantly provided with it, that even those most difficult to please in all other matters do not commonly desire more of it than they already possess. It is unlikely that this is an error on their part; it seems rather to be evidence in support of the view that the power of forming a good judgment and of distinguishing the true from the false, which is properly speaking what is called Good sense or Reason, is by nature equal in all men.[4]

Almost the same set of equivalent expressions are used in both the *Regulae* and the *Discourse* for *bona mens*, but there is the obvious difference that in the *Regulae* it is something which is to be sought after, while in the *Discourse* good sense is complete and equal in all men. In the *Regulae* the expression is equated with wisdom, while in the *Discourse* it plainly is not, for it cannot be said that all men are equally wise. The connection between the two, however, is made evident enough in the *Discourse*: "For to be possessed of good mental powers is not enough; the principal matter is to apply them well." It is *bona mens*, conceived as a natural faculty but applying itself with method, which gives rise to *bona mens*, conceived as wisdom, the highest perfection of our nature. The original title of the *Discourse on Method—Le projet d'une Science universelle qui puisse élever notre nature à son plus haut degré de perfection*—marks clearly the rôle attributed to method, or the "universal science," in the acquisition of wisdom.[5]

There is one essential characteristic of wisdom, its universality, which is a direct function of the natural faculty of good sense or reason. If there is the possibility of a wisdom that can judge of all matters that come before us, one of its conditions lies in the fact that the natural light of reason, considered as a native power of the mind, remains identical with itself no matter on what it is directed. It can be

---

[3]For a discussion of the translation into French of *bona mens*, and the translation into Latin of *bon sens*, see E. Gilson, *Discours de la Méthode, texte et commentaire* (Paris, 1925), 81 f.

[4]H. R., I, 81.

[5]"La sagesse n'est que le bon sens parvenu au point de perfection le plus haut dont il soit susceptible, grâce à la méthode qui n'en est elle-même que l'usage régulier." Gilson, *Comm.*, 82.

said of this natural faculty, as it is said of wisdom, that it "always remains one and the same, however applied to different subjects, and suffers no more differentiation proceeding from them than the light of the sun experiences from the variety of things it illuminates." Thus while the sciences may be many by virtue of their different subject-matters, they are one by virtue of the one and only exercise of the mind by which these subjects are known. There is only one mode of cognition, a direct seeing, a mental vision, or intuition, which remains the same for all its objects.

Intuition is defined by Descartes as "the undoubting conception of an unclouded and attentive mind, and springs from the light of reason alone."[6] It is a completely passive apprehension of its object. All actions of the mind must be attributed to the will; in knowing the mind does nothing.[7] A striking description of the passivity of the mind in intuition is given in a letter to the Marquis of Newcastle: "Intuitive knowledge is an illumination of the mind, by which it sees in the light of God[8] the things it has pleased him to reveal to it by a direct impression of the divine light upon our understanding, which in this respect is not considered as an agent, but only as receiving the rays of divinity." Thus, for example, in the absolute assurance of the proposition, *Cogito ergo sum*, we have knowledge which is not the result of any reasoning nor of any instruction: "your mind sees it, feels it, handles it."[9] In this letter Descartes distinguishes between intuition and reasoning, but in the *Regulae*, where he makes the same distinction, he at once reduces reasoning, in so far as it is cognitive, to intuition. It is true that inference involves an "action of a mind," by which it takes in successively the links connecting one truth with another. There must, however, be "a clear vision of each step in the process."[10] Reasoning is a temporal series of intuitions, each of which is a perfectly passive apprehension of a necessary connection between objects. The "actions" of the mind—and all actions must be referred

[6]*Reg.* III, H. R., I, 7.

[7]"It seems to me that it is . . . a passion of the soul to receive this or that idea, and that it is only its volitions that are actions." *Letter to P. Mesland,* 2 May, 1644(?), *Descartes, Correspondance,* ed. C. Adam and G. Milhaud (Paris, 1936 et seq.), VI, 142.

[8]More usually referred to by Descartes as the light of reason.

[9]March or April, 1648, *Œuvres de Descartes,* ed. C. Adam and P. Tannery (Paris, 1897-1913), V, 136–8.

[10]*Reg.,* III, H. R., I, 8.

to the will—are successive acts of attention, for looking is necessary for seeing.

There is, moreover, no diversity of cognitive faculties in the mind. Of the four faculties, understanding, imagination, sense, and memory, the latter three belong strictly to the body.[11] Only the understanding belongs to the mind.

It is a single agency, whether it receives impressions from the common sense simultaneously with the fancy, or applies itself to those that are preserved in the memory, or forms new ones. . . . It is one and the same agency which, when applying itself along with the imagination to the common sense, is said to see, touch, etc.; if applying itself to the imagination in so far as that is endowed with diverse impressions it is said to remember; if it turn to the imagination in order to create fresh impressions, it is said to imagine or conceive; finally, if it act alone it is said to understand. . . . Now it is the same faculty that in correspondence with those various functions is called either pure understanding, or imagination, or memory, or sense.[12]

This sole cognitive function of the mind of apprehending its objects by "an act of thought that is similar, single, and distinct,"[13] admits of no degrees.[14] Nor is this single cognitive agency possessed in different degrees by different persons. "Good sense or Reason is by nature equal in all men." One man may have a quicker mind than others, a more accurate and distinct imagination, or a more comprehensive memory. But "as to reason or sense . . . I would fain believe that it is complete in each individual."[15]

Rule I indicates two quite different relations which knowledge, or science, has to wisdom. On the one hand we are told that the sciences are to be pursued not for the sake of knowledge as such, nor for the

[11]Ibid., XII, H. R., I, 36–8.

[12]Ibid., 39. To Lord Herbert of Cherbury's plurality of cognitive faculties Descartes opposed a single faculty: "He [Lord Herbert] would have it that there are as many faculties in us as there are diversities of things to be known, which I can only understand to be like saying that, because wax can receive an infinity of shapes, it has within it an infinity of faculties for receiving them. In a sense this is true, but I do not see that anything useful can be had from such a way of speaking. . . . That is why I prefer to conceive that wax, by virtue of its flexibility alone, receives all kinds of shapes, and that the mind acquires all its knowledge by its reflection, either upon itself for intellectual objects, or upon the various dispositions of the brain, to which it is joined, for corporeal objects, whether these dispositions depend on the senses or on other causes. . . ." Letter to Mersenne, 16 Oct., 1639, A. M., III, 255.

[13]Ibid., IX, H. R., I, 29.

[14]"We must not fancy that one kind of knowledge is more obscure than another, since all knowledge is of the same nature throughout." Ibid., XII, H. R., I, 47.

[15]Disc., I, H. R., I, 82.

pleasure which comes from the sheer contemplation of truth, but because in pursuing them we are able to form that power of judgment or good sense which constitutes wisdom. This is the part of Descartes' teaching which was emphasized by the Port Royal logicians and was incorporated in the first "Discourse" of their *Art of Thinking*. They write,

Thus, the main object of our attention should be to form our judgment, and render it as exact as possible; and to this end, the greater part of our studies ought to tend. We apply reason as an instrument for acquiring the sciences; whereas, on the contrary, we ought to avail ourselves of the sciences, as an instrument for perfecting our reason—justness of mind being infinitely more important than all the speculative knowledges which we can obtain, by means of sciences the most solid and well-established. This ought to lead wise men to engage in these only so far as they contribute to that end, and to make them the exercise only, and not the occupation, of their mental powers. If we have not this end in view, the study of the speculative sciences, such as geometry, astronomy, and physics, will be little less than a vain amusement, and scarcely better than the ignorance of these things. . . .[16]

This is the conception of a wisdom devoid of content. It consists only in the perfection of a faculty of the mind through the study of the sciences, but not in the mind's possession of science. On the other hand, because some knowledge is logically dependent on other knowledge, Descartes indicates also that it is only in the actual possession of antecedent knowledge that the mind can exercise its wisdom in judging of all things. From this point of view science does not provide merely a training in wisdom, but is contained in wisdom. Thus in the latter part of his discussion of Rule I he moves from a consideration of the unity which the sciences possess by virtue of the unitary nature of cognition to a consideration of the unity which they possess by virtue of their logical interconnections. This is a unity, not of the mind in knowing, but of what is known by the mind.[17]

[16]*Logic, or the Art of Thinking* (2nd. ed., 1644) tr. T. S. Baynes (Edinburgh, 1850), 1 f.
[17]It may be noted that the logical unity of the sciences has here the status of something which we ought to "believe." The suppositional character of this unity is indicated also in the *Discourse on Method*, in which Descartes says, "Those long chains of reasoning, simple and easy as they are, of which geometricians make use in order to arrive at the most difficult demonstrations, *had caused me to imagine* that all those things which fall under the cognizance of man might *very likely* be mutually related in the same fashion." H. R., I, 92 (italics mine). In the *Search*

In the *Search after Truth by the Light of Nature* (ca. 1647?) it is the logical unity of the sciences which Descartes regards as making wisdom possible, for by traversing the logical connections linking together *all* possible knowledge, the mind is enabled to extend its range to include everything we may ever need to know.

The dialogue opens with the author's avowedly startling claim that he will be able to show how an individual can by his own efforts come to know, not only everything which is necessary for the direction of his life, but also everything, however curious, that comes within the limits of the human mind's capacity to know. "But . . . I warn you that what I undertake is not as difficult as might be imagined. Those branches of knowledge which do not extend beyond the capacities of the human mind are, as a matter of fact, united by a bond so marvellous, they are capable of being deduced from one another by sequences so necessary, that it is not essential to possess much art or address in order to discover them, provided that by commencing with those that are most simple we learn gradually to raise ourselves to the most sublime."[18] Although, as we have noted, Descartes' wisdom possesses a content of knowledge, it nevertheless requires very little of it and makes an absolutely minimum demand upon the use of memory. Because of the logical unity of the sciences, if the mind should actually possess only the principles of knowledge, it will potentially possess all the knowledge of which it is capable. Wisdom requires no encyclopaedic erudition, either in humanistic studies or in the sciences. The pursuit of learning in letters and in history, which was taken in the Renaissance as the very basis of wisdom, is condemned by Descartes as "folly" and a mere cramming of the memory. "I shall believe myself to have sufficiently fulfilled my promise if, in

---

after Truth, however, this logical unity is dogmatically asserted. H. R., I, 306, 327. Supposition occupies an important rôle in Descartes' theory of method, and is specified in the third of the rules given in the *Discourse*. Its use in physics is discussed below. The theory of knowledge given in *Reg.* XII, and which is also discussed below, is presented as suppositional; and as such is justified by Descartes on methodological grounds. H. R., I, 36. Supposition also has a methodological function in metaphysics. In *Meditation I* Descartes does not merely doubt the existence of external things but asserts positively what he has no grounds for believing to be true. "I shall suppose . . . some evil genius . . . etc." In his *Reply* to the *Objections IV* he comments, "For the analytic style of composition which I adopted allows us sometimes to make certain assumptions without their being as yet sufficiently investigated, as was evident in the first Meditation, in which I provisionally assumed many doctrines which I afterwards refuted." H. R., II, 116 f. See also H. R., II, 206.
18H. R., I, 306.

explaining to you the truths which may be deduced from common things known to each one of us, I make you capable of discovering all the others when it pleases you to take the trouble to seek them."[19]

Turning to the Preface to the *Principles of Philosophy*, we find that Descartes' definition of wisdom now includes within it the knowledge of principles.

This word philosophy signifies the study of wisdom, and . . . by wisdom we not only understand prudence in affairs, but also a perfect knowledge of all things that a man can know, both for the conduct of his life and for the conservation of his health and the invention of all the arts; and that in order that this knowledge should subserve these ends, it is essential that it should be derived from first causes, so that in order to study to acquire it (which is properly termed philosophising) we must begin with the investigation of these first causes, i.e. of the Principles. . . . It is really only God alone who has Perfect Wisdom, that is to say, who has a complete knowledge of all things; but it may be said that men have more wisdom or less according as they have more or less knowledge of the most important truths [i.e., of principles].[20]

UNITY OF METHOD

In Rule I Descartes takes the fact that the nature of knowing is the same for all subjects as implying that there is one method for all the sciences. And this is given, even prior to the fact of the logical interdependence of all knowledge, as a reason for not pursuing the different sciences in isolation from one another. Because of the transferability of method from one subject-matter to an entirely different one, success in the discovery of truth in one science is an aid to discovery of it in another science. One of the avowed aims of the *Discourse on Method* was to demonstrate in actual practice the universal applicability of the method. He explains that the three treatises appended to the *Discourse*, i.e., the *Dioptrics*, the *Meteors*, and the *Geometry*, had been called *Essays in this Method* to support his claim that what they contained could not have been discovered without that one method. "I have also inserted something of metaphysics, of physics, and of medicine in the first *Discourse* in order to show that it extends to every kind of subject-matter."[21]

That the unitary nature of the mind in knowing should imply

[19]*Ibid.*, 309 f.   [20]*Ibid.*, 204.
[21]*Letter to Mersenne*, 27 Feb., 1637, A. M., I, 329.

unity of method for all the sciences is made evident when Descartes later in the *Regulae* takes up the question of how we are able to determine the nature of method. This itself is a problem requiring solution by method. We must go about it in the same way as a mechanical craftsman who lacks the instruments of his craft. The smith, for example, will have to use whatever nature provides, such as a lump of iron as an anvil, a rock as a hammer, and pieces of wood as tongs. But the first thing he will do with these is not to make swords and helmets, but to fashion the tools which he needs for making swords and helmets. And so also with the sciences; we can use certain rough precepts of method, which seem to be innate to the mind, not, however, in order to apply them immediately to solving the problems of the sciences, but to the problem of determining what is necessary for discovering the truth. Therefore, the very first inquiry to be undertaken by means of our rough precepts of method should be "that which seeks to determine the nature and scope of human knowledge," because in the pursuit of this investigation, "the true instruments of knowledge and *the whole method of inquiry come to light*."[22] This is the procedure which Descartes himself then applies for eliciting the method of the sciences. His theory of method has its foundations in theory of knowledge.

The first thing to be done in the pursuit of the problem of the nature of knowledge is to divide it into two parts, "for it ought either to relate to us who are capable of knowledge, or to the things themselves which can be known: and these two factors we discuss separately."[23] On the side of ourselves who know, there are the four faculties, understanding, imagination, sense, and memory. It has already been seen that of these the mind itself possesses only the faculty of understanding. The other three are faculties of the body, of which the mind can make use in exercising its one and only function. They are its "instruments." Turning to the second part of the problem, "the objects themselves which are to be known," we find that Descartes' account of them is entirely determined by the theory of the nature of the mind given in his answer to the first part of the problem, for these objects, he says, "must be considered only in so far as they are objects of the understanding."[24] It is only in terms of that relation that they are relevant to the problem. From this

[22]*Reg.* VIII, H. R., I, 26.   [23]*Ibid.*, 27.
[24]*Ibid.*

point of view the primary division of things is into "the class (1) of those whose nature is of the extremest simplicity and (2) of the complex and composite."[25] These are purely epistemological simples and compounds, which are not to be confused with physical simples and the things compounded out of them.

. . . relatively to our knowledge single things should be taken in an order different from that in which we should regard them when considered in their more real nature. Thus, for example, if we consider a body as having extension and figure, we shall indeed admit that from the point of view of the thing itself it is one and simple. For we cannot from that point of view regard it as compounded of corporeal nature, extension and figure, since these elements have never existed in isolation from each other. But relatively to our understanding we call it a compound constructed out of these three natures, because we have thought of them separately before we were able to judge that all three were found in one and the same subject. Hence here we shall treat of things only in relation to our understanding's awareness of them, and shall call those only simple, the cognition of which is so clear and so distinct that they cannot be analysed by the mind into others more distinctly known. Such are figure, extension, motion, etc.; all others we conceive to be in some way compounded out of these.[26]

The basis for this primary division of the objects of knowledge into simple natures and compounds of simple natures is the theory that intuition is the sole cognitive exercise of the mind. The first kind of object appropriate to intuition is something so simple that it cannot be analysed into anything more distinctly known. Once having apprehended these simples the mind can then extend the same direct vision to any necessary connections between them, and in that way come to know compounds of simple natures. Beyond that it cannot go. "No knowledge is at any time possible of anything beyond those simple natures and what may be called their intermixture or combination with each other."[27] Although the act of compounding is referred to deduction, it is, however, "quite clear," says Descartes, "that this mental vision extends both to all those simple natures and to the knowledge of the necessary connections between them."[28] The number of the simple natures on which all science rests is regarded by Descartes as extremely limited.[29]

[25]Ibid.
[27]Ibid., 45.
[26]Ibid., XII, H. R., I, 40 f.
[28]Ibid., 43.
[29]"There are but few pure and simple natures, which either our experiences or some sort of light innate in us enable us to behold as primary and existing per se, not as depending on any others." Ibid., VI, H. R., I, 16.

Once having made his primary division of the objects of knowledge, Descartes then introduces a second division, by which simple natures are distinguished according to their content—some are purely intellectual, some are purely material, and some are common to both intellect and matter indifferently, like existence, unity, duration, etc. To the latter may also be added certain common notions which govern inferences and which therefore serve to connect together the other simple natures. It is only in this second division that Descartes introduces such diversities in the simple natures as would correspond to any differences of subject-matter in the sciences. The nature of method is not affected, however, by this second division. It is already fully determined by the primary one. This is made plain if we consider the nature of the relation which method has to knowables. "In order to know these simple natures," says Descartes, "no pains need be taken, because they are of themselves sufficiently well known. Application comes in only in isolating them from each other and scrutinizing them separately with steadfast gaze."[30] Once they have been secured by analysis, they need only to be combined for our knowledge to extend itself. This provides the basis for Descartes' definition of method.

Method consists entirely in the order and disposition of the objects towards which our mental vision must be directed if we would find out any truth. We shall comply with it exactly if we reduce involved and obscure propositions step by step to those that are simpler, and then starting with the intuitive apprehension of all those that are absolutely simple, attempt to ascend to the knowledge of all others by precisely similar steps. In this alone lies the sum of all human endeavour. . . .[31]

It will be irrelevant to the operations of analysis and combination whether the simple natures which are the termini of our analyses are intellectual or material. The method of both metaphysics and physics will remain the same.

THE LOGICAL ORDERING OF THE SCIENCES

Not only, however, is the method of all the sciences the same, but method knows no lines of division separating the different sciences according to their subject-matters. To Mersenne Descartes writes,

[30]*Ibid.*, XII, H. R., I, 46.   [31]*Ibid.*, V, H. R., I, 14.

It is to be observed in everything I write that I do not follow the order of subject-matters, but only that of reasons, that is to say, I do not undertake to say in one and the same place everything which belongs to a subject, because it would be impossible for me to prove it satisfactorily, there being some reasons which have to be drawn from much remoter sources than others; but in reasoning by order, *a facilioribus ad difficiliora*, I deduce thereby what I can, sometimes for one matter, sometimes for another, which is in my view the true way of finding and explaining the truth; and as for the ordering of subject matters, it is good only for those for whom all reasons are detached, and who can say as much about one difficulty as about another.[32]

Since all objects of knowledge are mutually related, the totality of scientific knowledge, organized according to the "order of reasons," will comprise a single science, or deductive system, resting on a small number of first causes or principles.

Although Descartes tells Mersenne that the ordering of knowledge according to subject-matters disrupts the ordering according to reasons, he does, nevertheless, distinguish sciences within the one comprehensive deductive system, for he says, "Thus philosophy as a whole is like a tree whose roots are metaphysics, whose trunk is physics, and whose branches, which issue from the trunk, are all the other sciences. These reduce themselves to three principal ones, viz., medicine, mechanics and morals—I mean the highest and most perfect moral science which, presupposing a complete knowledge of the other sciences, is the last degree of wisdom."[33] This mode of distinguishing the sciences is, however, not inconsistent "with the order of reasons," but is directly based on it. "Metaphysics" is an expression employed by Descartes in two senses. Taken in one of these senses, it is the science of "immaterial or metaphysical things" and as such stands contrasted with physics, the science of "corporeal or physical things." Thus in the dedicatory letter prefixed to the *Meditations* (1641) metaphysics is given as having for its object the knowledge of God and the soul by natural reason. Metaphysics is also, however, the science of "the principles of knowledge," and it is as taken in the latter sense that metaphysics, or first philosophy, is capable of being distinguished as a science from all the others in his single deductive system. As the science of principles metaphysics is not confined to immaterial substances. Its subject is "all the clear and

[32]*Letter to Mersenne*, 24 Dec., 1640, A. M., IV, 239.
[33]*Principles of Philosophy*, Preface, H. R., I, 211.

simple notions which are in us."[34] Included among these is the primitive notion of matter. One of the most important of his intentions in the *Meditations on the First Philosophy* was to establish the distinction between the primitive notions of mind and body, or thought and extension. If, however, we take metaphysics not as the science of principles, but as the science of immaterial things, i.e., of God and the soul, it is impossible to render it complete prior to physics, for some of it depends upon deductions from physics—for example, the proof for the immortality of the soul. Descartes explains that this proof has not been included in his *Meditations on the First Philosophy* "because the premises from which the immortality of the soul may be deduced depend on an elucidation of a complete system of Physics."[35] This is an outstanding instance of the "order of reasons" running counter to the "order of subject-matters."

Considered as the science of the principles of knowledge metaphysics supplies the logical foundations of physics. To Mersenne Descartes writes, "I shall tell you, between ourselves, that these six *Meditations* contain the entire foundations of my physics."[36] What is it in the *Meditations* which is sufficient to lay the entire foundations of his physics? It is simply the clear and distinct idea of matter as extension in length, breadth, and depth, with its absolute distinction from the idea of thinking substance.[37] It is this which lays the ghost of substantial forms and disposes of the physics of the scholastics.

In the tree of knowledge the branches issuing from the trunk, which is physics, are the three practical sciences, medicine, mechanics, and morals—and it is ultimately for the sake of these that philosophy exists. In the *Discourse on Method* Descartes recounts his discovery of the revolutionary import of his new physics: that instead of being a mere speculative philosophy like that taught in the Schools, it was indeed a practical philosophy, providing us with the means by which

---

[34]"The principal aim of my Metaphysics is to explain what those things are which we can distinctly conceive." *Letter to Mersenne*, 30 Sept., 1640, A. M., IV, 170.

[35]*Meditations*, Synopsis, H. R., I, 141.

[36]Descartes continues, "But it is not necessary to say so, if you please; for those who favour Aristotle would perhaps experience more difficulty in approving them; and I hope that those who read them will become insensibly accustomed to my principles and recognize their truth before realizing that they destroy those of Aristotle." *Letter to Mersenne*, 28 Jan., 1641, A. M., IV, 269.

[37]For a fuller account of Descartes' conception of metaphysics as the science of the principles of knowledge see the Appendix.

"to render ourselves the masters and possessors of nature."[38] There were three spheres in which he foresaw this control of nature being exercised: in the mechanical arts, in medicine, and in morality. But, however prodigious the possibilities to be opened up in the first of these spheres—here his anticipations were not less than Bacon's— it was the mastery of nature as it exists in man's body which concerned him most, not only as a means to the preservation of physical health, but also for morality, for the passions have their causes in the body. The study of the machinery of the purely animal organism would make possible the acquisition of "a very absolute dominion over the passions" by even the weakest minds.[39] Thus "the order of reasons" allows for the distinguishing within the single deductive system of a science of principles and a science of what is deduced from these principles. It is physics which is deduced. But physics makes possible the mastery and possession of nature. Since there are three ends which can be served by this mastery, the application of physics to each of these ends gives rise to a distinction of three practical sciences, medicine, mechanics, and morals.

THE UNIVERSAL SCIENCE AND THE METHOD OF MATHEMATICS

What does Descartes' conception of a universal science owe to mathematics?[40] At the time of writing the *Regulae* he considered that as yet men possessed only two sciences, arithmetic and geometry, for they alone possessed the certainty that is the mark of science. They therefore constituted a model for all other kinds of scientific inquiry, in the sense that "in our search for the direct road towards truth we should busy ourselves with no object about which we cannot attain a certitude equal to that of the demonstrations of Arithmetic and

[38]H. R., I, 119.

[39]"Since we can with a little industry change the movement of the brain in animals deprived of reason, it is evident we can do so yet more in the case of men, and that even those who have the feeblest souls can acquire a very absolute dominion over all their passions if sufficient industry is applied in training and guiding them." *The Passions of the Soul*, I, 50, H. R., I, 356. At the time that he was composing the *Passions* Descartes wrote to Chanut, "I shall tell you in confidence that the notion of physics, which I have sought to acquire, has been enormously useful to me just as it is for establishing the certain foundations of morals." 15 June, 1646, A. T., IV, 441.

[40]We are not concerned here with psychological influences. These belong to speculative biography.

Geometry."[41] At the same time Descartes was very critical of existing mathematics. His criticisms are made on the score of *method*. Thus existing mathematics provides the paradigm of certainty in the sciences but not of method in the sciences. In the *Regulae* he recounts that when he first took up mathematics he read nearly everything that mathematicians had written, particularly in arithmetic and geometry, "because they were said to be the simplest and so to speak the way to all the rest. But in neither case did I then meet with authors who fully satisfied me. I did indeed learn in their works many propositions about numbers which I found on calculation to be true. As to figures, they in a sense exhibited to my ideas a great number of truths and drew conclusions from certain consequences. But they did not seem to make it sufficiently plain to the mind itself why these things are so, and how they discovered them."[42] To this he adds a reference to "those superficial demonstrations which are discovered more by chance than by skill." In both the *Regulae* and the *Discourse* he criticizes the analysis of the ancients and the algebra of the moderns as methods respectively of geometry and arithmetic, and it was as methods that he ranked them together with the logic of the syllogism. He had, he said, studied all three because they "seemed as though they ought to contribute something to the design I had in view." That design was to find "the true method of arriving at a knowledge of all things of which the mind is capable."[43] But after finding each defective in certain respects, he concludes, "This made me feel that *some other Method* must be found, which, comprising the advantages of the three, is yet exempt from their faults." (Italics mine.) Descartes thereupon presents his famous four rules of method, which he believed he should find "quite sufficient," provided that he always steadfastly adhered to them.

That something of the true method is present in existing mathematics Descartes acknowledges. "I am quite ready to believe that the greater minds of former ages had some knowledge of it, nature even conducting them to it. For the human mind has in it something of the divine, wherein are scattered the first germs of useful modes of thought." He finds evidence of the existence of this natural endowment in the geometrical analysis of the ancients, though they ignobly

[41]*Reg.* II, H. R., I, 5.   [42]*Ibid.*, IV, H. R., I, 11.
[43]*Disc.* II, H. R., I, 91.

suppressed their knowledge, and in that "kind of Arithmetic called Algebra, which designs to effect, when dealing with numbers, what the ancients achieved in the matter of figures." But, he adds, "nothing is less in my mind than ordinary Mathematics. . . . I am expounding quite another science, of which these illustrations are rather the outer husks than the constituents. Such a science should contain the primary rudiments of human reason, and its province ought to extend to the eliciting of true results in every subject"—i.e., not just in numbers and figures.[44]

Reflection upon these imperfections in the methods of mathematics took him from the particular studies of arithmetic and geometry to a general investigation of mathematics. What, he asked, is meant by "mathematics"—a term which embraces not only arithmetic and geometry, but also astronomy, music, optics, mechanics, and several other sciences? These are all called parts of mathematics. Descartes concluded that all subjects were referred to mathematics in so far as they involved the investigation of "order and measurement," whether it be in numbers, figures, stars, sounds, etc. "I saw consequently that there must be some general science to explain that element as a whole which gives rise to problems about order and measurement, restricted as these are to no special subject matter. This, I perceived, was called 'Universal Mathematics' . . . because in this science is contained everything on account of which the others are called parts of Mathematics."[45] Descartes' experience is recounted again more briefly in the *Discourse* where he describes it as his earliest effort in the application of his new method. He noted that although all the particular mathematical sciences have different objects, they all agree in being concerned only with "the various relations or proportions which are present in these objects." Therefore he confined himself to these proportions taken in their most general aspect, considering them only in those objects which made the knowledge of them easiest, without, however, restricting them to these objects.

Descartes' first attempt, then, in the application of his new method was in mathematics, and this attempt appears to have had two very important consequences for him. In the first place, there was confirmation of the validity of his method in the sheer fact of the success that attended its use—"the exact observation of the few precepts I

[44]*Reg.* IV, H. R., I, 11.   [45]*Ibid.*, 13.

had chosen gave me so much facility in sifting out all the questions embraced in these two sciences [geometrical analysis and algebra] that in the two or three months which I employed in examining them ... not only did I arrive at the solution of many questions which I had hitherto regarded as most difficult, but, towards the end, it seemed to me that I was able to determine in the case of those of which I was still ignorant, by what means, and in how far, it was possible to solve them."[46] Just because his method was not a specifically mathematical one, but an absolutely general one, the success of its application in mathematics gave promise that it could be applied equally well in any other science.[47]

In Rule XIV of the *Regulae* Descartes gives a more detailed account of how the method of mathematics was brought under a method more general than itself. Universal mathematics is the science of "order and measurement," while the method common to all the sciences is the science of "order." He points out that differences of relation or proportion hold either between numerical quantities or between continuous magnitudes. "Order" has reference particularly to the former, and "measure" to the latter. Relations of continuous magnitudes can, however, be reduced, at least in part, to expressions of relations between numbers by means of an assumed or imputed unity. "Further it is possible to arrange our assemblage of units in such an order that the problem which previously was one requiring the solution of a question in measurement, is now a matter merely involving an inspection of order,"[48] i.e., it has been subsumed under the universal science of order. And, therefore, Descartes at once points out that it is his method which has made possible the progress which this transformation produces, for it is his method which reduces the solutions of all problems of any kind into the discovery of "order."[49]

[46]*Disc.* II, H. R., I, 93 f.
[47]He says, "not having restricted this method to any particular matter, I promised myself to apply it as usefully to the difficulties of other sciences as I had done to those of algebra." *Ibid.*
[48]*Reg.* XIV, H. R., I, 64.
[49]Descartes here makes the same claim for his universal method in relation to mathematics as Leibniz was later to make for his. Leibniz's general science or method, the *art of combinations*, he defined as "the science of forms or of similarity and dissimilarity," while algebra he defined as "the science of magnitude, or of equality and inequality." "Taken by itself algebra has only rules of equality and proportion, but when the problems are more difficult and the roots of equations very involved, algebra is forced to draw something again from a higher science of similitude and

The second important consequence which emerged from Descartes' application of his method to mathematics is contained in his observation that while mathematics itself embraces a large number of subordinate sciences, yet the diversity of their subject-matters introduces no element of diversity into their method. Universal mathematics provides a concrete demonstration of the way in which method can be indifferent to subject-matters, and in its relation to the particular mathematical sciences, universal mathematics becomes a kind of model for an absolutely universal science or method, bearing an analogous relation to all sciences of any kind. In all of them, that element by virtue of which they are sciences will be simply the investigation of problems of order, in no matter what kind of objects they arise. To say, however, that universal mathematics provides a model for a universal science or method is not to say that Descartes borrowed his conception of this method from mathematics. It is to say, rather, that he constructed the particular model in the light of his universal science. The method was not derived from mathematics but applied to mathematics.

Mathematics, however, occupies a role of the greatest importance in the acquisition of method considered as a *skill*.[50] "The employment of the rules which I here unfold is much easier in the study of Arithmetic and Geometry (and it is all that is needed in learning them) than in inquiries of any other kind. Furthermore, its usefulness as a means towards the attainment of a profounder knowledge is so great, that I have no hesitation in saying that it was not the case that this part of our method was invented for the purpose of dealing with mathematical problems, but rather that mathematics should be studied solely for the purpose of training us in this method."[51] He himself had studied mathematics with a view to no practical results except that his mind should "become accustomed to the nourishment

dissimilitude or from the science of combinations." This device of solving equations by reducing them to similar equations or equations of the same form was, he says, used by Cardan and Vieta. "But this art can be and ought to be used not only when our concern is with formulas which express magnitudes, and with the solution of equations, but also when the involved key is to be developed for other formulas which have nothing in common with magnitude." *Letter to Tschirnhaus*, May, 1678, *Gottfried Wilhelm Leibniz, Philosophical Papers and Letters*, ed. L. E. Loemker (Chicago, 1956), 295.

[50]He says, "it consists more in practice than in theory." *Letter to Mersenne*, 27 Feb., 1637, A. M. I, 329.

[51]*Reg.* XIV, H. R., I, 57.

of truth." In commenting to Burman on this statement in the *Discourse* Descartes explains, "anyone who has once accustomed his mind to mathematical reasoning will keep it apt for the inquiry into other truths, for reasoning is everywhere identical."[52] The mathematical sciences deal with the simplest of all objects; they are therefore "the easiest and clearest" of all the sciences. A skill is acquired by exercising it first upon the simplest objects. Once acquired it can be extended to more difficult ones. For that reason mathematics is an essential propaedeutic to the study of all the other sciences. In the order of a man's self-instruction, after providing himself with a provisional code of morals, "he should . . . study logic—not that of the Schools . . . but the logic that teaches us how best to direct our reason in order to discover those truths of which we are ignorant. And since this is very dependent on custom, it is good for him to practise the rules for a long time on easy and simple questions such as those of mathematics. Then when he has acquired a certain skill in discovering the truth in these questions, he should begin seriously to apply himself to the true philosophy," i.e., metaphysics, physics, medicine, mechanics, and morals, in their deductive order.[53]

PHYSICS AS BROUGHT UNDER THE UNIVERSAL SCIENCE

In considering the relation of mathematics to Descartes' method, it becomes necessary, then, to distinguish two separate questions: first, how we come to know what the true method is—this is revealed in theory of knowledge—and, secondly, how we can accustom the mind to the use of the method. It is evident that Descartes himself does not regard mathematics as the foundation of the first but only of the second. The confusion of the two is responsible for a widely held view that Descartes conceived his method as a mere generalization of the method of mathematics. This in turn has led some to think that Descartes must have believed that the whole of physics could be deduced *a priori* from self-evident principles, and without any reference to experience. Since, however, a small acquaintance with his writings reveals that he placed great emphasis on experience and was himself actively engaged as an experimenter, this is sufficient to

[52]*Entretien avec Burman*, A. T., V, 176.
[53]*Principles*, Preface, H. R. I, 211.

dispel such a version of his thought. But once his emphasis on experience is taken into account, it is then used in evidence against him. For, it is argued, to claim to be using the *a priori* and deductive method of mathematics in physics, and then to insist on the necessity of experience is to do nothing less than to confess that the method is a failure.[54]

The question to be asked, however, in assessing this alleged confession of failure is not whether experience can have any place in mathematical reasoning, but whether it has any place in the method of science as actually formulated by Descartes. His definitive formulation is given in the four rules in the *Discourse on Method*, of which "the third was to carry on my reflections in due order, commencing with objects that were the most simple and easy to understand, in order to rise little by little, or by degrees, to knowledge of the most complex, assuming an order, even if a fictitious one, among those which do not follow a natural sequence relatively to one another."[55]

The reference to *supposition* in this rule carries us back to the fuller discussion of it which is contained in the *Regulae*. In Rule X Descartes recommends the study of crafts, like weaving and tapestry, and of number games, in order to accustom the mind to the discovery of order, before we go on to deal with more serious and difficult problems in the sciences. He says,

It was for this reason that we insisted that method must be employed in studying those matters; and this in those arts of less importance consists wholly in the close observation of the order which is found in the object studied, whether that be an order existing in the thing itself, or due to subtle human devising. Thus if we wish to make out some writing in which the meaning is disguised by the use of a cypher, though the order here fails to present itself, we yet make up an imaginary one, for the purpose both of testing all the conjectures we may make about single letters, words or sentences, and in order to arrange them so that when we sum them up we shall be able to tell all the inferences that we can deduce from them.[56]

There can have been no intention on Descartes' part to confine the rôle of hypotheses or "suppositions" to cyphers and humanly devised riddles (as Gilson and Kemp Smith assert). Otherwise, there would

[54]Cf. Leon Roth, *Descartes' Discourse on Method* (Oxford, 1937), chapter v.
[55]The original reads as follows for the last phrase: "et supposant même de l'ordre entre eux qui ne précèdent point naturellement les uns les autres."
[56]H. R., I, 31.

have been no point in including the use of them in the four rules of method for the sciences, as given in the *Discourse on Method*. Indeed, Descartes, in the concluding sections of his *Principles of Philosophy*, as Leibniz was also to do, describes the discovery of the explanations of observable phenomena as analogous to the unlocking of a cypher. The use of hypotheses is necessary in both. There is no reason to suppose that Descartes used the comparison with any less precise an intention than did Leibniz, who adopted almost verbally this section of Descartes' *Principles* and frequently reiterated this analogy with cryptography.[57]

The nature of order in all science is the same whether it is hypothetical or not. Physics is as deductive as mathematics, but in physics, in so far as it provides explanations of *particular* phenomena, the causes from which effects are deduced are hypothetical. These hypotheses stand or fall according as the deduced effects agree or disagree with experience. Hence although a physical theory is deductive, it is "proved" by experience. Descartes acknowledges in the *Discourse* that *a posteriori* proof of a deductive theory appears to involve a logical circle, for in deduction it is the causes which demonstrate the effects, while in the case under consideration it is also said that the effects reciprocally demonstrate the causes. But the circularity is an illusion arising from an ambiguity in the word "to demonstrate." To his correspondent Morin, who makes the charge of circularity, Descartes writes, "There is a great difference between *proving* and *explaining*. To which I add that the word *to demonstrate* can be used to signify one or the other, at least if it is taken according to common usage, and not with the particular meaning given to it by philosophers." Descartes then repeats what he has said in the *Discourse* about his *Dioptrics* and *Meteors* as hypotheses, namely, "since experience renders the greater part of these effects very certain, the causes from which I deduce them do not so much serve to prove them as to explain them; but it is the causes which are proved by the effects."[58]

Descartes also explains in the *Discourse*, however, that he had called the *Dioptrics* and *Meteors* "hypotheses" because he had not

[57]Of Leibniz's use of it Couturat says, "Cette comparaison de la méthode des sciences physiques à l'art de déchiffrer n'est ni fortuite ni paradoxale: elle repose sur l'analogie réelle des deux méthodes." *La Logique de Leibniz* (Paris, 1901), 268. See further chapter IV, 76, of the present book.
[58]13 July, 1638, A. M., II, 311.

yet given an *a priori* proof for the suppositions with which they began, but he believed himself to be capable of deducing them from the primary truths of metaphysics. In the meantime the two treatises were presented as sufficiently established *a posteriori* by experience. After the publication of his *Meditations*, however, Descartes was able to say that "these six *Meditations* contain the entire foundations of my physics" or "contain all its principles."[59] What was promised in the *Discourse* was now accomplished.

Does this mean that metaphysics now supplants the necessity for all *a posteriori* proof in physics? When we look at the *Principles of Philosophy* in which the three books of Descartes' physics follow upon the first one containing his metaphysics, it becomes clear that the grounding of physics in metaphysics does not mean the elimination of all hypothesis or suppositional order from physics and therefore of *a posteriori* proof. At the conclusion of this work he writes,

But here it may be said that although I have shown how all natural things can be formed, we have no right to conclude on this account that they were produced by these causes. For just as there may be two clocks made by the same workman, which though they indicate the time equally well and are externally in all respects similar, yet in nowise resemble one another in the composition of their wheels, so doubtless there is an infinity of different ways in which all things that we see could be formed by the great Artificer (without it being possible for the mind of man to be aware of which of these means he has chosen to employ). This I most freely admit; and I believe that I have done all that is required of me if the causes I have assigned are such that they correspond to all the phenomena manifested by nature (without inquiring whether it is by their means or by others that they are produced). And it will be sufficient for the usages of life to know such causes, for medicine and mechanics and in general all these arts to which the knowledge of physics subserves, have for their end only those effects which are sensible, and which are accordingly to be reckoned among the phenomena of nature.

But nevertheless, that I may not injure the truth, we must consider (two kinds of uncertainty and) first of all what has moral certainty; that is, a certainty which suffices for the conduct of life, though if we regard the absolute power of God, what is morally certain may be uncertain. . . . If, for instance, anyone wishing to read a letter written in Latin characters that are not placed in their proper order, takes it into his head to read B wherever he finds A and C where he finds B, thus substituting for each letter the one following it in the alphabet, and if he in this way finds that there are certain Latin words composed of these, he will not doubt that

[59]*Letters to Mersenne*, 28 Jan., 1641, 11 Nov., 1640, A. M., IV, 269, 200.

the true meaning of the writing is contained in these words, though he may discover this by conjecture, and although it is possible that the writer did not arrange the letters in this order of succession, but on some other, and thus concealed another meaning in it: for this is so unlikely to occur (especially when the cipher contains many words) that it seems incredible. But they who observe how many things regarding the magnet, fire, and the fabric of the whole world, are here deduced from a very small number of principles, although they considered that I had taken up these principles at random and without good grounds, they will yet acknowledge that it could hardly happen that so much would be coherent if they were false.[60]

But Descartes then adds that there are "some even among natural things, which we judge to be absolutely, and more than morally, certain."[61] In other words, there are in physics some elements of explanation which are not hypothetical. It is these which are established *a priori* by metaphysics. What the *Meditations* show, thereby laying the foundations of physics, is that matter consists only of extension, and can possess no properties except the modes of extension, namely, size, figure, situation, and local motion. Everything which follows from this or, as Descartes says, everything which is "derived in a continual series from the first and most simple principles of human knowledge," can be known with certainty.[62] Thus metaphysics establishes *a priori* that all explanations of natural phenomena must be mechanical. But because there is always the possibility of different mechanical explanations for what is observed, just as there may be different mechanisms within the two clocks which are outwardly in every respect similar, it follows that the explanations of particular phenomena remain hypothetical. What are not hypothetical are the *a priori* general principles of mechanics in terms of which all admissible hypotheses must be constructed. They have an absolute certainty which is established in metaphysics.

[60]*Principles* IV, cciv, ccv, H. R., I, 300 f.
[61]*Ibid.*, ccvi, H. R., I, 301.
[62]*Ibid.*, 302.

IV. LEIBNIZ:
THE DEMONSTRATIVE
ENCYCLOPAEDIA

IN TWO SHORT DOCUMENTS, *Precepts for Advancing the Sciences and Arts* (1680) and *Discourse touching the Method of Certitude, and the Art of Discovery in order to End Disputes and to Make Progress Quickly* (ca. 1680), Leibniz presented his main arguments of persuasion to enlist the powerful support of Louis XIV for a scheme to create a "demonstrative encyclopaedia."[1] He sought to show, in the first place, that the critical state of learning and the sciences made this project of most pressing urgency, and secondly, that it was actually possible to give to knowledge of all kinds a *demonstrative* form. These documents bring together some of the dominant themes, which are repeated and developed in many of his scattered and fragmentary writings.

Although Leibniz was to point with satisfaction to the fact that civilization had now reached full maturity and that many first-rate minds were enthusiastically pursuing scientific inquiry, thus opening up the possibilities in the immediate future of extraordinary advances in the sciences, he did not—surprisingly, if we consider his reputation for optimism—regard this progress as inevitable. On the contrary he considered the times to be exceedingly precarious. The existing condition of learning contained within it the seeds of its own dissolution, the loss of all its gains slowly accumulated through the centuries,

---

[1] "Ce projet a occupé Leibniz pendant toute sa vie; ce devait être sa grande œuvre philosophique et scientifique. Aussi l'histoire de cette entreprise, de ses origines, de ses transformations, de son avortement final, se confond-elle avec l'histoire de la pensée et de l'esprit du philosophe." L. Couturat, *La Logique de Leibniz* (Paris, 1901), 119. Chapter v contains a detailed history of the project.

the abandonment of the arts and sciences—"the true treasure of mankind"—and the lapse into apathy, ignorance, and barbarism. The threat to civilization arose from the total want of co-operation in the efforts of scientists and in the chaotic amassing of man's intellectual possessions in "that horrible mass of books which keeps on growing." He found that scientists were working at cross-purposes and with ill will towards one another. Instead of trying to decrease their disputes, they were trying to increase them. If this condition should be allowed to persist, it could result in a general disgust with the sciences and the abandonment of hope for the achievement of anything valuable. For science to fall into disrepute would have the fatal consequence of man's reverting to barbarism. At the same time the haphazard multiplying of books would ultimately result in their not being read at all and their authors being lost to human memory. To be an author would then cease to be honourable and it could even become disgraceful. The motive to contribute to knowledge would vanish and only a literature of light entertainment would remain.

For Leibniz the making of a demonstrative encyclopaedia provided the sole means of avoiding these disastrous possibilities. Such an encyclopaedia would in the first place serve to co-ordinate the efforts of all scientists in a single enterprise of total comprehension, and by its very nature as demonstrative it would put an end to all controversies. The method by which the sciences were to be rendered demonstrative would leave them as little subject to disagreement as arithmetic and geometry and all disputes would be resolved by calculations as precise as those characterizing these sciences. Secondly, the encyclopaedia would prevent the loss of the knowledge slowly accumulated in the past, on the continued use of which present civilization rests—a treasure lying dispersed in an increasing chaos of books. "Why refer to some remote posterity what would be incomparably easier in our own day, since the confusion has not yet risen to the point it will reach then? Could any century be more suited than ours which will be recorded some day as the century of discoveries and wonders?"[2]

But there was also knowledge which had never even been recorded, and Leibniz was convinced that this wealth exceeded both in quantity

[2]*Precepts for Advancing the Sciences and Arts*; *Leibniz, Selections*, ed. P. P. Wiener (New York, 1951), 32.

and in importance anything to be found in books. Some of it was private and would be lost with its owners. There were certain specialized skills, not easily communicated, which, though seemingly trivial, could have important implications for other human activities. "Hunters, fishermen, merchants, sea voyagers, and even games of skill as well as of chance, furnish material with which to augment useful sciences considerably. Even in the games of children there are things to interest the greatest Mathematician."[3] For example, the compass-needle and the air-gun had had their origins in children's amusements. Hence nothing in the practical life of men should be considered too trifling for record. In order to indicate the enormous scope of what he was proposing, Leibniz writes

. . . just imagine how much knowledge one would need if suddenly transported to a desert island one had to make for himself everything useful and convenient which the abundance of a big city furnishes us, since the city is filled with the best workmen and most talented men from all social ranks. Or else, imagine that an art is lost and that it must be rediscovered; all our libraries could not help supply the art, for though I do not disagree that there are a great many admirable things in books which men in the professions are still ignorant of and should take advantage of, it is nevertheless a fact that the most important observations and turns of skill in all sorts of trades and professions are as yet unwritten. This fact is proved by experience when passing from theory to practice we desire to accomplish something.[4]

THE REDUCTION OF KNOWLEDGE TO PRINCIPLES

Descartes, in describing the nature of universal wisdom, had maintained that if we actually possess only the principles of science, and know how to make the necessary deductions, we potentially possess all that we may ever wish to know as occasion arises. The mind would thus be left free from an encumbering mass of detail. The same notion underlies Leibniz's device for securing the sum of human knowledge, both recorded and unrecorded, from the risk of oblivion. He would have all his collaborators contribute to the encyclopaedia by finding the principles of their respective sciences. When these were taken in conjunction with "the General Science or art of

[3] *Discourse Touching the Method of Certitude, and the Art of Discovery in order to End Disputes and to Make Progress Quickly*, Wiener, 47.
[4] *Ibid.*, 48.

discovery," they would be sufficient to enable the whole of the content of these sciences to be recovered whenever they should be required.

Concerning the principles of scientific discovery, it is important to consider that each science usually consists of some few propositions which are either observations of experiment or veins of thought which have offered the occasion and means for discovery. They would suffice to recover the discovery if it were lost and to learn it without a teacher if one wished to apply himself enough, by combining those few propositions in the usual way with the precepts of a higher science, assumed to be already known, namely, either the general science or art of discovery, or another science to which the science in question is subordinate. For example, there are several sciences subordinate to Geometry in which it is enough to be a geometer and to be informed of a few leading facts or principles of discovery to which geometry may be applied, so that it is not necessary in addition to discover for one's self the principal laws of these sciences."[5]

Leibniz cites as examples the theory of perspective, the theory of the sundial which is only a corollary of a combination of astronomy and perspective, and music which is subordinate to arithmetic. Given a few fundamental experiments with harmonies and dissonances, all the other general precepts of music depend on arithmetic. It would then be possible to show a man knowing nothing about music how to compose without mistakes. This would not mean, of course, that he could be shown how to compose beautiful music, but here Leibniz willingly acknowledges a limitation upon the extent to which an art is capable of being reduced to principles and transmitted in an encyclopaedia, and he is even anxious to prevent any facile misinterpretation of his claims. It is theory alone in an art, not practice, which can be reduced to principles. But the distinction between theory and practice is one which he finds generally to be misunderstood. In art there are some things which depend on imagination and the spontaneity of genius. Just as playing the clavichord requires the acquisition of a certain habit by the fingers, so imagining a beautiful melody, writing a good poem, or spontaneously sketching architectural ornaments requires the acquisition by the imagination of certain habits. Here precisely is where we are dependent on practice rather than on theory. The acquired skills of the imagination lie outside the scope of a demonstrative encyclopaedia. But apart from these there are things in the arts in which we can succeed merely by the use of

[5]*Precepts*, Wiener, 41

reason aided by a few experiments and observations. "In all matters where it is possible for judgment aided by a few precepts to avoid application and experiment, we can always reduce all of a science with its subordinate parts to a few fundamentals or principles of discovery sufficient to determine all the questions which can arise in the circumstances by combining with the principles the exact method of the true Logic or art of discovery."[6]

The demonstrative encyclopaedia would, then, reduce all acquired knowledge to manageable proportions; it would leave the mind unburdened by a mass of learning; and it would enable all the sciences and the theoretical part of the arts to be recovered in their entirety should Europe be overwhelmed by barbarism, provided only that the encyclopaedia should survive. But it was much more than a mere preservation of the treasure of mankind that concerned Leibniz. The encyclopaedia was an essential instrument of the progress of the sciences and it opened up the possibility of their immeasurable advance in the future, and of the perfection of civilization, provided that advantage were taken of the opportunity arising at this precise stage in their development.

THE POSSIBILITY OF RENDERING ALL KNOWLEDGE, EVEN OF THE PROBABLE, DEMONSTRATIVE

To win royal support Leibniz also felt compelled to show that his plan was not a fanciful one by providing some evidence that all branches of knowledge were actually capable of demonstration in the strict sense in which that term is used in geometry. The possibility of such demonstration rested on "the General Science or art of discovery." Leibniz found it inopportune, however, to reveal "this great artifice" in the proposals to be laid before Louis XIV, and excused himself from doing so. Nevertheless he was able to point to some areas outside of geometry in which geometrical reasoning had already been successfully applied. In philosophy itself, where this rigorous reasoning was most needed, several attempts had been made to organize arguments in Euclidean form, notably by Descartes at the end of his *Reply* to the second set of *Objections* and by Spinoza in his *Principles of Descartes' Philosophy* and in his *Ethics*, although

[6]*Ibid.*, 44.

Leibniz could find little resemblance between these and geometrical reasoning "except in outer garb."[7] A striking instance of success did exist, however, in jurisprudence, and Leibniz advanced it as "an odd but true paradox, that there are no authors whose manner of writing resembles the style of the Geometers more than the style of the Roman jurisconsults whose fragments are found in the Pandects. . . . they put our philosophers to shame in even the most philosophical matters which they are often obliged to treat. There is no excuse, therefore, in philosophy to claim it is impossible to preserve the required exactitude of reasoning."[8] Jurisprudence had performed an important generative rôle in the development of Leibniz's schemes for a demonstrative encyclopaedia. It was in connection with his early concern with the law that he had first formed the idea of taking action to save "the public treasure of learning" from the threat of disgust and reversion to barbarism arising from the disorderly accumulation of books and the confusion of studies. All these materials were to receive a systematic and logical ordering in accordance with the principles of didactic method which he had worked out for the law in his *New Method of Learning and Teaching Jurisprudence* of 1667.[9]

But the greatest promise of the possibility of rendering all knowledge demonstrative could be found in that area which by popular opinion was the most recalcitrant of all to demonstration, namely knowledge of the merely probable. It was the mathematicians who had shown the possibility of exact demonstrations of degrees of probability once the impetus had been provided by the Chevalier de Méré, the gambler, when he put his questions on bets to Pascal. Huygens had taken these questions up in his book on Chance, and others had followed. The Pensioner de Witt had then put their

[7]"There has appeared a metaphysical Euclid by Thomas White, and Abdias Trew, a skilful mathematician of Altdorf, has reduced Aristotle's Physics to a demonstrative form as much as that author was able, and Father Fabry has claimed he has dressed up all philosophy as geometry." *Ibid.*, 37. Among others whom Leibniz mentions as having attempted to use demonstration outside of mathematics are Galileo in the science of motion, Kepler, Gilbert, Snell in dioptrics, Hobbes in ethics and physics, and Sir Kenelm Digby in proving the soul's immortality; see *The Art of Discovery* (1685), Wiener, 52–6.

[8]*Precepts*, Wiener, 38. In the project for legal reform presented in his early *Elements of Natural Law* (1670–1), Leibniz points to the obviously demonstrative character of theory of law, for it is a science resting solely on definitions, "and definition is the principle of demonstration." *Gottfried Wilhelm Liebniz, Philosophical Papers and Letters,* ed. L. E. Loemker (Chicago, 1956), 206 f.

[9]See Couturat, *La Logique de Leibniz*, 123 f.

principles to work in his treatise on annuities. Here was evidence that even in the probable it is possible to deduce exactly from the given premisses what is most probable.

A new logic of probability, supplanting the Topics of Aristotle, was to be included in Leibniz's General Science or art of discovery. It would in the first place be of the utmost importance in morals and politics. The principles of the theoretical part of these sciences are purely *a priori*, and possess apodictic certainty, but, as Leibniz points out, "to apply them to practice, we need a new kind of logic entirely different from that which we have had up until now. It is this which has been principally lacking in the practical sciences."[10] Another important field for the logic of probability was in physics, not, however, in this case, because it is a practical science—though its applications are practical, and as such require a logic of probability —but because it involves experience and hypotheses. The foundations of physics are not themselves hypothetical for Leibniz, and no logic of probability is required to establish them. It is possible to determine *a priori* that all explanations of material phenomena must be mechanical. Moreover it is possible and, indeed, necessary to demonstrate the general laws of mechanics by showing that they follow from still higher principles belonging to the science of metaphysics.[11] But, while the most general principles of explanation are *a priori* and certain, it is impossible for the human mind to determine *a priori* the causes of particular phenomena, for contingent facts require an infinite analysis for arriving at the grounds of their necessity. In seeking their causes the human mind is able to proceed only by hypotheses or conjectures. Nevertheless physics remains deductive. No "science" can consist of empirical generalizations.[12] Leibniz's conception of the hypothetico-deductive method in physics is very close to Descartes', but to it he adds his logic of probability for the evaluation of an hypothesis.

[10]Letter to Burnett, 17/27 July, 1696, *Die philosophischen Schriften von Gottfried Wilhelm Leibniz*, ed. C. I. Gerhardt (Berlin, 1875–90), III, 183.

[11]*New Essays Concerning Human Understanding* (1704), IV, xii, 13, tr. A. G. Langley (La Salle, 1949), 527; *On the Elements of Natural Science* (ca. 1682–4), Loemker, 447; *Critical Thoughts on the General Part of the Principles of Descartes* (1692), on Part II, art. 64, Loemker, 674–6.

[12]"There are in fact *experiments* which succeed countless times in ordinary circumstances, yet instances are found in some extraordinary cases in which the experiment does not succeed . . . the senses and induction can never teach us truths that are fully universal or absolutely necessary, but only what is and what is found in particular examples." *Letter to Queen Sophia Charlotte of Prussia*, 1702, Loemker, 895.

The conjectural method *a priori* proceeds by hypotheses, assuming certain causes, perhaps, without proof, and showing that the things which now happen would follow from these assumptions. A hypothesis of this kind is like the key to a cryptograph, and, the simpler it is, and the greater the number of events that can be explained by it, the more probable it is. But just as it is possible to write a letter intentionally so that it can be understood by means of several different keys, of which only one is the true one, so the same effect can have several causes. Hence no firm demonstration can be made from the success of hypotheses. Yet I shall not deny that the number of phenomena which are happily explained by a given hypothesis may be so great that it must be taken as morally certain.[13]

The degree of probability to be attached to an hypothesis is, he says in this passage, determined by the number of events that can be explained by it. In a letter to Conring he adds two other factors: simplicity or paucity of assumptions, and power of prediction: " . . . a hypothesis becomes the more probable as it is simpler to understand and wider in force and power, that is, the greater the number of phenomena that can be explained by it, and the fewer the further assumptions. . . . Those hypotheses deserve the highest praise (next to truth), however, by whose aid predictions can be made, even about phenomena or observations which have never been tested before; for a hypothesis of this kind can be applied, in the realm of practice, instead of truth."[14]

THE GENERAL SCIENCE

It is the General Science which will render all sciences demonstrative, but what is this General Science? In some, but certainly not all, of its most fundamental aspects it bears a strong resemblance to Descartes' universal method. Indeed, in a paper entitled *On Wisdom*, Leibniz, without mentioning his name, takes Descartes' rules of method as a starting point. The essential core of the method is to be found in those rules concerning "analysis" and "combination or synthesis."

The *art of discovery* consists of the following maxims:

1. In order to become acquainted with a thing we must consider all

[13]*On the Elements of Natural Science*, Loemker, 437, cf. Descartes, *Principles of Philosophy, The Philosophical Works of Descartes*, tr. E. S. Haldane and G. R. T. Ross (Cambridge, I, 1931; II, 1934), IV, ccv. See chapter III, 67, of the present volume.

[14]*Letter to Herman Conring*, 19 March, 1678, Loemker, 288.

of its prerequisites, that is, everything which suffices to distinguish it from any other thing. This is what is called definition, nature, essential property.

2. After we have found a means of distinguishing it from every other thing, we must apply this same rule to the consideration of each condition or prerequisite entering into this means, and consider all the prerequisites of each prerequisite. And that is what I call *true analysis*, or distribution of the difficulty into several parts.

3. When we have pushed the analysis to the end, that is, when we have considered the prerequisites entering into the consideration of the proposed thing, and even the prerequisites of the prerequisites, and finally have come to considering a few natures understood only by themselves without prerequisites and needing nothing outside themselves to be conceived, then we have arrived at a *perfect knowledge* of the proposed thing. . . .

9. The fruit of several analyses of different particular matters will be the catalogue of simple thoughts, or those which are not very far from being simple.

10. Having the catalogue of simple thoughts, we shall be ready to begin again *a priori* to explain the origin of things starting from their source in a perfect order and from a combination or synthesis which is absolutely complete. And that is all our soul can do in its present state.[15]

It was in connection with Leibniz's study of formal logic that the possibility of "a catalogue of simple thoughts," out of which all other concepts could be derived by combination, first came to him.

As a boy I learned logic, and, having already developed the habit of digging more deeply into the reasons for what I was taught, I raised the following question with my teachers. Seeing that there are categories for the simple terms by which concepts are ordered, why should there not also be categories for complex terms, by which truths may be ordered? I was then unaware that geometricians do this very thing when they demonstrate and order propositions according to their dependence upon each other. It seemed to me, however, that this could be achieved universally if we first had the true categories for simple terms, and if, to obtain these, we set up something new in the nature of an alphabet of thoughts, or a catalogue of the highest genera, or of those we assume to be highest, such as a, b, c, d, e, f, out of whose combination inferior concepts may be formed.[16]

[15]*On Wisdom* (ca. 1693), Wiener, 78–80. The Cartesian theory that "no knowledge is at any time possible of anything beyond those simple natures and what may be called their intermixture or combination with each other," gives rise to the two moments of his method: the art of discovering the simple constituents of knowledge by analysis, and then the art of combining them, or synthesis.

[16]*On Universal Synthesis and Analysis, or the Art of Discovery and Judgment* (1679?), Loemker, 351.

It was, he says, the realization that if we possessed this alphabet everything else could be discovered, which was the occasion of his youthful work, *The Art of Combinations* (1666). He describes his idea as one of "astounding import," and, indeed, its consequences were to be far reaching for all his later thought on the method and unity of the sciences.

In accordance with this doctrine all concepts will be either primary or derivative. Of clear concepts those are primary which are understood through themselves alone and are incapable of further analysis. Derivative concepts are the results of combining primitive concepts, or of combining less composite into more composite concepts. This provides the basis for Leibniz's theory of definition by which there are two ways in which a thing can be defined. It can be given a *nominal* definition, which consists only in the enumeration of such characteristics as are sufficient for recognizing the thing. Or it can be given a *real* definition, which consists in an analysis of a concept into its components, so that when the analysis is complete the concept can be seen to contain no contradiction. In other words, a real definition demonstrates the possibility of the concept.

Definitions are the sole foundation of all demonstration. Nor are axioms an exception to this statement, for they themselves are demonstrable.

From such ideas or definitions . . . there can be demonstrated all truths with the exception of identical propositions, which by their very nature are indemonstrable and can truly be called axioms. What are popularly called axioms, however, can be reduced to identities by analysing either the subject or predicate or both, and so demonstrated; for by assuming the contrary, we can show that the thing would at the same time be and not be. . . . Thus any truth whatever can be justified, for the connection of the predicate with the subject is either evident in itself, as in identities, or can be explained by an analysis of terms. This is the only, and the highest criterion of truth in abstract things, that is, things which do not depend on experience—that it must either be an identity or be reducible to identities.

From this can be derived the elements of eternal truth in all things in so far as we understand them, as well as the method for proceeding demonstratively, as in geometry.[17]

In this passage the principle that all true propositions are either identities or reducible to identities is restricted to the sciences which

[17]*Ibid.*, 356.

do not depend on experience. But in fact, as we soon see, it is extended even to empirical propositions, and in its universal application it receives the name, *principle of sufficient reason*. This is made clear in *First Truths* (ca. 1680–4) in which Leibniz writes,

First truths are those which predicate something of itself or deny the opposite of its opposite. . . .

All other truths are reduced to first truths with the aid of definitions or by the analysis of concepts; in this consists *proof a priori*, which is independent of experience. . . .

The predicate or consequent therefore always inheres in the subject or antecedent. . . . In identities this connection and the inclusion of the predicate in the subject are explicit; in all other propositions they are implied and must be revealed through the analysis of the concepts, which constitutes a demonstration *a priori*.

This is true, moreover, in every affirmative truth, universal or singular, necessary or contingent, whether its terms are intrinsic or extrinsic denominations. Here lies hidden a wonderful secret which contains the nature of contingency or the essential distinction between necessary and contingent truths. . . .

These matters have not been adequately considered because they are too easy. . . . At once they give rise to the accepted axiom that *there is nothing without a reason, or no effect without a cause*. Otherwise there would be a truth which could not be proved *a priori*, or resolved into identities—contrary to the nature of truth, which is always either expressly or implicitly identical.[18]

Contingent propositions differ from necessary propositions only in that their truth requires an infinite analysis in order to establish identity, and such an infinite analysis is impossible for the human mind. Contingency is a term, therefore, which has significance only in relation to the human mind. For God, who comprehends the infinite, all true propositions are necessary in the same way as the truths of the mathematical sciences are for us. This limitation upon the human mind's power of analysis is hinted at in Leibniz's *Meditations on Knowledge, Truth and Ideas* (1684), where he remarks, "Whether men will ever be able to carry out a perfect analysis of concepts, that is, to reduce their thoughts to the *first possibles* or to irreducible concepts, or (what is the same thing) to the absolute attributes of God themselves or the first causes and the final ends of things, I shall not now venture to decide."[19] In his treatment of con-

[18]*First Truths*, Loemker, 411–13.
[19]*Meditations on Knowledge, Truth, and Ideas*, Loemker, 452.

tingent truths, however, he has shown that this completed analysis is not in fact possible in the case of the sciences involving experience, though it is possible in the purely *a priori* sciences. Thus the alphabet of human thoughts which he sought is only partially attainable, and for the remainder of the alphabet we must be content with incompletely analysed concepts. These, however, like the ultimately irreducible thoughts, will also be the results of analysis, and when taken together with them will comprise "the catalogue of simple thoughts, or those which are not very far from being simple." With these "we shall be ready to begin again *a priori* to explain the origin of things starting from their source in a perfect order and from a combination or synthesis which is absolutely complete. And that is all our soul can do in its present state."[20]

THE UNIVERSAL CHARACTERISTIC

Such being the nature of demonstration, why is it that up until now it has been found only in mathematics? This is a question to which Leibniz directs himself in his *Preface to the General Science* (1677). He finds the answer in the fact that mathematics carries its own test with it; that is to say, the tests or experiments which verify the reasoning are not performed on the things themselves, but on the characters which have been substituted for them. These are experiments which cost only paper and ink and they reveal the slightest errors. Thus, if in order to determine the result of multiplying 1677 by 365, we had to make 365 piles of 1677 pebbles and count them all, we could hardly expect ever to arrive at the correct answer. But on paper the operation is done easily and accurately. Or if a certain value is assigned to π we do not have to construct a material circle and to measure it with a string in order to discover whether the ratio of the length of the string to the diameter has the asserted value. To determine an error of one-thousandth (or less) part of the diameter would involve all the practical difficulties of constructing a huge circle with extreme accuracy. But if the experiment is done on paper using characters in place of the things themselves, a false value is easily detected.

Whence it is manifest that if we could find characters or signs appropriate for expressing all our thoughts as definitely and as exactly as

[20]*On Wisdom*, Wiener, 80.

arithmetic expresses numbers or geometric analysis expresses lines, we could in all subjects *in so far as they are amenable to reasoning* accomplish what is done in Arithmetic and Geometry.

For all inquiries which depend on reasoning would be performed by the transposition of characters and by a kind of calculus, which would immediately facilitate the discovery of beautiful results. For we should not have to break our heads as much as is necessary today, and yet we should be sure of accomplishing everything the given facts allow.

Moreover, we should be able to convince the world what we should have found or concluded, since it would be easy to verify the calculation either by doing it over or by trying tests similar to that of casting out nines in arithmetic. And if someone would doubt my results, I should say to him: "Let us calculate, Sir," and thus by taking to pen and ink, we should soon settle the question."[21]

Leibniz's reference to the limits of reasoning is to indicate that certain experiments are always necessary to serve as a basis for reasoning. Nevertheless, once we have the experiments we can, through the instrumentality of characters, derive from these experiments all that it is possible to derive and also we can see where further experiments need to be performed if anything still remains doubtful. In political science and medicine, where it is impossible to have all the experimental data necessary for passing infallible judgments, we shall still be able to determine with mathematical accuracy what, on the basis of the data we do have, is most probable.

Once we have the characters for expressing all our thoughts, we shall be in possession of a new language, one which will be as easily spoken as written. Because of the facility with which it can be learned and its great usefulness it will rapidly attain universal adoption. "This language will be the greatest instrument of reason. I dare say that this is the highest effort of the human mind, and when the project will be accomplished it will simply be up to men to be happy since they will have an instrument which will exalt reason no less than what the Telescope does to perfect our vision."[22] Nothing can contribute more to the general welfare of mankind than the perfection of reason.

When the catalogue of human thought has been achieved, and each concept has been given its characteristic number or sign, we shall have secured a complete demonstrative encyclopaedia of all knowledge.

[21]*Preface to the General Science* (1677), Wiener, 15.
[22]*Ibid.*, 16.

Now since all human knowledge can be expressed by the letters of the Alphabet, and since we may say that whoever understands the use of the alphabet knows everything, it follows that we can calculate the number of truths which men are able to express, and that we can determine the size of a work which would contain all possible human knowledge, in which there would be everything which could ever be known, written, or discovered; and even more than that, for it would contain not only the true but also the false propositions which we can assert, and even expressions which signify nothing.[23]

Though Leibniz grants that men will always be content with something less than all the knowledge of which they are capable, nevertheless, theoretically, if they kept on steadily progressing, in the end they would exhaust all possible knowledge. After that no novel could be written which had not already been written, no new dream would be possible, nor could anything be said which had not already been said.

Although Leibniz believed that the universal characteristic could be very easily learned—he suggested on one occasion that it would take only a few days[24]—nevertheless it would be very difficult to construct. And although he never did succeed in constructing it, in 1679 he expressed his belief "that a few selected persons might be able to do the whole thing in five years, and that they will in any case after only two years arrive at the doctrines most needed in practical life, namely, the propositions of morals and metaphysics, according to an infallible method of calculation."[25]

The highly organized co-operation of others was required for the construction of the characteristic, and it is easy to see why this was so. The work of analysis necessary for reaching the elements of knowledge which were to be symbolized would have to be undertaken throughout all the sciences. "We must believe that only gradually and by diverse efforts or by the work of several persons shall we come to those demonstrative elements of all human knowledge."[26] Nothing less was required, he says, than the founding of "a mathematical-philosophical course of study" in order to establish the characteristic numbers for all ideas. It was in order to get this necessary co-operation of the learned that Leibniz had to look to royal powers of command.

[23]*The Horizon of Human Doctrine* (after 1690), Wiener, 75.
[24]*Letter to John Frederick, Duke of Brunswick-Hanover*, Autumn, 1679, Loemker, 402.
[25]*Towards a Universal Characteristic* (1677), Wiener, 22.
[26]*Precepts*, Wiener, 39.

From this it is clear that while the characteristic was a necessary condition for establishing the demonstrative encyclopaedia, the encyclopaedia was at the same time necessary for establishing the universal characteristic. As Couturat has pointed out, the two projects had to be developed in mutual conjunction.[27]

THE DISSOLUTION OF ALL DIVISIONS OF THE SCIENCES

In his successive plans for the encyclopaedia Leibniz drew up several tentative systems of classification of the sciences, but by the time he came to write his detailed commentary on Locke's *Essay Concerning Human Understanding* these had ceased to have any logical significance. One of the remarkable consequences for the individual sciences of the construction of the demonstrative encyclopaedia would be the dissolving of all divisions among them. The last chapter of Locke's *Essay* entitled "Of the Divisions of the Sciences" provided Leibniz with the occasion for making this point. Locke had adopted the ancient classification of the sciences into physics, ethics, and logic.[28] In his *New Essays Concerning Human Understanding* (1704) Leibniz paraphrased Locke's statement in the following way:

All that can enter into the sphere of human understanding is either the nature of things in themselves, or in the second place, man in the character of an agent, tending towards his end and in particular towards his happiness, or in the third place the means of acquiring and communicating knowledge. *Science* then is divided into three kinds. The first is *Physics* or Natural Philosophy, which comprises not only bodies and their properties, as number, figure, but also spirits, God himself and the angels. The second is *Practical Philosophy* or *Ethics*, which teaches the means of obtaining good and useful things, and proposes to itself not only the knowledge of the truth, but also the practice of that which is right. Finally, the third is *Logic* or the knowledge of the signs, for λόγος signifies word. We need *signs* of our ideas to enable us to communicate our thoughts to one another, as well as to register them for our own use. Perhaps if we should consider distinctly and with all possible care that this last kind of science revolves about ideas and words, we should have a logic and criticism different from that which has hitherto been seen.

[27]*La Logique de Leibniz*, 79 f.
[28]"Philosophic doctrine, say the Stoics, falls into three parts: one physical, another ethical, and the third logical. Zeno of Citium was the first to make this division in his *Exposition of Doctrine. . . .*" Diogenes Laertius, *Lives of Eminent Philosophers*, tr. R. D. Hicks (London, 1950), VII, 39.

And these three kinds, Physics, Ethics, and Logic, are like three great provinces in the intellectual world, entirely separate and distinct the one from the other.[29]

Leibniz notes that like the ancients Locke has included under logic "all that relates to words and the explication of our thoughts: *Artes dicendi*," and he completely dissociates himself from this conception of logic. For him it is "the science of reasoning, of judgment, of invention." But the principal difficulty he found in this division was that each part seemed to include the whole of knowledge. For example, the doctrine of spirits, i.e., substances possessing understanding and will, which belongs to physics, will involve a thorough inquiry into the nature of the understanding. It will therefore have to absorb the whole of logic. Then in dealing with the nature of the will it must inquire into good and evil, happiness and misery, and in the end will include all practical philosophy. On the other hand, practical philosophy can include everything, for everything is relevant to our happiness. As for logic, as concerned with language, it cannot avoid invading the sciences, for the sciences can only be treated by giving definitions to their terms. "Here, then, your three great provinces of encyclopaedia are in continual war, since one is always encroaching upon the rights of the others."[30] All classifications of the sciences are arbitrary.[31] The same truth can be put in different places, according to the terms it contains, or according to its causes, or according to the consequences which can be drawn from it. If we try to classify by the terms, then we find that a simple categoric proposition has two terms, a hypothetic proposition may have four, and there are still other more complex propositions. Thus "one and the same truth may have many places according to the different relations it can have."[32]

Instead, therefore, of taking the ancient division of physics, ethics, and logic as one separating three different sciences, Leibniz takes them as corresponding to three ways in which the whole of human knowledge, *taken as a whole*, can be disposed. The first way is to rank truths "according to the order of proofs, as the mathematicians

[29]*New Essays*, IV, xxi, Langley, 621 f.
[30]*Ibid.*, 623.
[31]"The entire body of the sciences may be regarded as an ocean, continuous everywhere and without a break or division, though men conceive parts in it and give them names according to their convenience." *The Horizon of Human Doctrine*, Wiener, 73.
[32]*New Essays*, Langley, 624.

do, so that each proposition would come after those on which it depends."[33] This is the *synthetic* or *theoretic* disposition of knowledge. But the same knowledge can be given a reverse order, by beginning with man's end, his happiness, and seeking the means for attaining it and for avoiding the evils contrary to it. This is the *analytic* or *practical* disposition of knowledge. This distinction between the two methods has, he points out, been employed in geometry. Euclid presented his geometry in the synthetic form, that is, as a science. But others have treated geometry not as a science but as an art and have used the analytic method. Both methods have their place in the encyclopaedia. In constructing the encyclopaedia, however, repetitions will have to be avoided through the use of references. In addition to the synthetic and analytic there is a third disposition of knowledge, that according to terms. This would be an index. Such an index could either arrange the terms according to predicaments common to all notions, or it could be alphabetical according to the languages used by scholars. "Now this index would be necessary in order to find together all the propositions into which the term enters in a sufficiently remarkable manner; for according to the two preceding ways, where truths are arranged according to origin or use, truths concerning one and the same term cannot be found together." Here Leibniz indicates how thoroughly the demonstrative ordering of knowledge destroys the classification of knowledge according to subject-matters. The index becomes necessary for bringing together all the propositions which bear on the same subject, and once again Leibniz makes a reference to geometry, which bears out his point that the demonstrative order does not permit everything belonging to the same subject to be dealt with in the same place. For that reason an index would also be of great utility in geometry.

Now considering these three dispositions, I find it remarkable that they correspond to the ancient division, which you have renewed, which divides science or philosophy into theoretic, practical and discursive, or rather into Physics, Ethics, and Logic. For the synthetic disposition corresponds to the theoretic, the analytical to the practical, and that of the index according to the terms to logic: so that this ancient division does very well, provided we understand these dispositions as I have just explained, i.e. not as distinct sciences, but as different arrangements of the same truths as far as we judge it advisable to repeat them.[34]

[33]*Ibid.*   [34]*Ibid.*, 625 f.

THE DEBT TO MATHEMATICS

The same question may be asked of Leibniz's conception of the unity of the sciences as was asked of Descartes'. What does it owe to mathematics? Leibniz's account of the origin of his idea of a general science—his early *Art of Combinations*—shows, as we have seen, that it was formed in his youth prior to any acquaintance with mathematics, when, he says, he "did not know that the geometricians do exactly what I was seeking when they arrange propositions in an order such that one is demonstrated from another."[35] But he also confesses that he would not have got very far with his idea without mathematics.

I consider it certain that the art of reasoning can be carried incomparably higher [than the existing logic] and believe not only that I see this but that I already have a foretaste of it, which I could hardly have attained, however, without mathematics. Though I found some basis for it even before I was a novice in mathematics, and had already printed something about it in my twentieth year, I have finally come to see how blocked are the ways to it and how hard it would have been to open them without the aid of the deeper mathematics.[36]

As he tells us, the chief thing to be learned from mathematics is that mathematics is demonstrative because it employs characters. From that fact alone he concludes "that if we could find characters or signs appropriate for expressing all our thoughts as definitely and as exactly as arithmetic expresses numbers or geometric analysis expresses lines, we could in all subjects *in so far as they are amenable to reasoning* accomplish what is done in Arithmetic and Geometry,"[37] i.e., in physics, metaphysics, ethics, politics, jurisprudence, and medicine. Nobody, so far, he says, "has gotten hold of a language which would embrace both the technique of discovering new propositions and their critical examination—a language whose signs or characters would play the same rôle as the signs of arithmetic for numbers and those of algebra for quantities in general. And yet it is as if God, when he bestowed these two sciences on mankind, wanted us to realize that our understanding conceals a far deeper secret, foreshadowed by these two sciences."[38] It is, then, from mathematics that Leibniz would

[35]*Towards a Universal Characteristic*, Wiener, 19.
[36]*Letter to Gabriel Wagner*, 1696, Loemker, 763.
[37]*Preface to the General Science*, Wiener, 15.
[38]*Towards a Universal Characteristic*, Wiener, 18.

claim to have extracted his secret, namely the rôle of signs or characters in making science possible.

Why, for Leibniz, does the substitution of signs or characters for things make reasoning and science possible? The answer is to be found in the abstractive function which he attributed to all *expression*.

That is said to express a thing in which there are relations (*habitudines*) which correspond to the relations of the thing expressed. But there are various kinds of expression; for example, the model of a machine expresses the machine itself, the projective delineation on a plane expresses a solid, speech expresses thoughts and truths, characters express numbers, and an algebraic equation expresses a circle or some other figure. What is common to all these expressions is that we can pass from a consideration of the relations in the expression to a knowledge of the corresponding properties of the thing expessed. Hence it is clearly not necessary for that which expresses to be similar to the thing expressed, if only a certain analogy is maintained between the relations.[39]

It is precisely because of the identity or analogy of the structure of relations in the thing used as symbol and the thing symbolized—two things which in other respects are totally dissimilar—that the symbol serves to abstract these relations and render them alone that which is expressed. But since it is solely these relations which are relevant to reasoning, the character or symbol, though its power of setting them free, makes reasoning possible.

In geometry the figures we draw must be regarded as characters, but, says Leibniz, a circle described on paper is not so useful a character as a number, because there is still some degree of similarity between the drawn figure and what it stands for, whereas a written number, say 10, has no similarity to what it stands for, nor 0 with nothing, nor *a* with a line. But if we possess well-invented characters, then we can introduce the same relation or order in the characters as there is in things.

I notice that, if characters can be applied to ratiocination, there is in them a kind of complex mutual relation (*situs*) or order which fits the things; if not in single words, at least in their combination and inflection, although it is even better if found in single words themselves. Though it varies, this order somehow corresponds in all languages. . . . For although characters are arbitrary, their use and connection have something which is not arbitrary, namely, a definite analogy between characters and things, and the relations which different characters expressing the same thing

[39]*What is an Idea?* (1678), Loemker, 318.

have to each other. This analogy or relation is the basis of truth. For the result is that, whether we apply one set of characters or another, the products will be the same, or equivalent or correspond analogously."[40]

In estimating the extent of Leibniz's debt to mathematics, we can say that prior to his acquaintance with mathematics he had definitely formed the idea of the alphabet of human thoughts and the conviction that all knowledge would consist in combinations of these elements. This meant that all ratiocination is concerned only with the relations by which these elements are combined. But it was from mathematics that he learned that it is only by substituting characters for the things which enter into the relations that the relations can be elicited for ratiocination to deal with. In that it does use characters for this purpose, mathematics shows itself to be merely a special instance of the use of a method of universal applicability.[41]

[40]*Dialogue* (1677), Loemker, 282.
[41]"Thus the best advantages of algebra are only samples of the art of characters whose use is not limited to numbers or magnitudes." *The Horizon of Human Knowledge*, Wiener, 74.

V. CONDILLAC:
THE ABRIDGEMENT OF
ALL KNOWLEDGE IN
"THE SAME IS THE SAME"

THE CRITIQUE OF SYSTEM MAKING

Like other philosophers of the French Enlightenment, Condillac set himself against the philosophy of Descartes and turned with admiration to Locke and Newton. To Locke he acknowledged his greatest debt, but not without criticism. Among other things Locke lacked "order."[1] Condillac's own thought is characterized by an extreme concern with the systematization and unity of knowledge. His attitude towards systematization was, nevertheless, in his own time the subject of misunderstanding. Though he himself did not misunderstand Condillac, d'Alembert was to remark that "the taste for systems . . . is today banished almost completely," and to Condillac he gave the credit for having delivered the final blows to *l'esprit de système*."[2] As the author of the *Traité des systèmes* (1749) Condillac was ironically destined to bear the reputation of being the principal enemy of *system* as such. On this undeserved reputation he commented sadly:

. . . a good *system* is only a well developed principle. I know that many people take this word in bad part, thinking that every *system* is a gratuitous hypothesis, or something worse like the dreams of the metaphysicians. . . . If I am told that I have written against *systems*, I beg people to read my

[1]*Cours d'études pour l'instruction du Prince de Parme; Histoire moderne*, XX, xii, *Œuvres philosophiques de Condillac*, ed. Georges Le Roy (Paris, I, 1947; II, 1948; III, 1951), II, 234a.
[2]*Discours préliminaire de l'Encyclopédie*, ed. F. Picavet (Paris, 1894), 116.

work right to the end, although it is humiliating for a writer to recognize that he is not read to the end. If at least they were to read the beginning carefully, they would see that I do not reject all *systems*.[3]

It was one of Condillac's reiterated maxims that "a well treated science is a well made system." Systematization he took to be an essential and inescapable feature of all thinking. It is not merely philosophers who systematize; everyone does, progressing naturally from prejudice to prejudice, from opinion to opinion, and from error to error. Nature compels us to make systems and for Condillac nature is the ultimate authority on method in the sciences.

The *Traité des systèmes* presents an exhaustive classification of the possible types of system. There are only three, and of these the third is the true or legitimate one. The difference between the three systems does not consist in the way in which their parts are connected, but in what they take for their principles or starting point, for in any system there are only two things—principles and consequences. The first kind of system starts with definitions and axioms and proceeds by the method of synthesis. These abstract principles even when they happen to be true and well-formed—though in metaphysics, ethics, and theology they are generally vague and badly formed—are improperly called principles for the simple reason that they are not what we know first. Merely to refer to them as abstract is to acknowledge that other things are known before them. They are useless for the making of discoveries since they are no more than abbreviated expressions for knowledge already acquired, and as such the most that they can do is to carry us back again to that knowledge. "In short," says Condillac, "they are maxims containing only what we know, and just as the people have their proverbs, these so-called principles are the proverbs of the philosophers, and that is all they are."[4]

The second is the hypothetical system, which adopts for principles certain suppositions which are designed to explain known phenomena. The suppositions are said to be *proved* by their felicity in *explaining*. Metaphysicians in particular have been extremely inventive in this kind of principle, and indeed there is no longer anything which remains mysterious for metaphysics, which they describe as "the science of first truths, or of the first principles of things." Condillac

[3]"Sistème [*sic*]," *Dictionnaire des synonymes* (found at Condillac's death), *Œuvres*, III, 511b f.
[4]*Traité des Systèmes*, xii, *Œuvres*, I, 195b.

does not deny that hypotheses can perform a legitimate rôle in scientific inquiry. Inquiry begins with provisional observations which give rise to suspicions or conjectures suggesting further observations to be made. But such conjectures cannot provide the logical foundations of a science.

Since suppositions are only suspicions, they are not established facts. Therefore they cannot be the principle or beginning of a system, otherwise the whole system would reduce itself to a suspicion.

But if they are not the principle or beginning of the system, they are the principle or beginning of our means for discovery of the system. Now, it is just because they are the principle of these means that they have also been taken to be the principle of the system. Thus two quite different things have been confused with one another.[5]

There is, nevertheless, one circumstance in which it is possible for a supposition to function, not merely as a means to the discovery of a first principle, but as itself an actual first principle. This occurs only when there is some way of establishing the truth of the supposition. For this two things are required. First, we must be able to exhaust all possible suppositions bearing on a question, and, secondly, we must be in possession of some means for determining our choice among them and of eliminating those which are false. In mathematics this is possible; in physics it is not. It was the taste for this second kind of system, resting on conjectures, which d'Alembert was to identify with *l'esprit de système*. No more than Condillac did he condemn all system, but he distinguished between what he called *le véritable esprit systématique* and *l'esprit de système* and warned that we must be careful not to take the one for the other.

The third and only valid system is that which takes experience for its starting point and proceeds by analysis.[6] This analysis forms the subject of a major part of Condillac's philosophical writings, for it is common to every science and to every inquiry he himself undertakes for the instruction of his royal pupil in his *Cours d'études* (1775):

[5] *Ibid.*, i, *Œuvres*, I, 123b.
[6] In contrasting the methods of synthesis and analysis it is necessary to take into account that for Condillac analysis contains two moments. There is first an act of decomposition which separates out the elements of an initially vague complex whole so that these can be distinctly perceived. This is followed by an act of recomposition which makes possible the simultaneous perception of the successively distinguished elements. "We decompose only in order to recompose. . . . Thought is not analysed until the complete whole is embraced at once." *La Logique*, I, ii, *Œuvres*, 376a.

". . . this method is the only one, and . . . it must be absolutely the same in all our investigations, for to study the different sciences is not to change the method, but to apply the same method to different objects."[7] Its unswerving uniformity is asserted with tireless repetition.

LANGUAGE AS THE INSTRUMENT OF ANALYSIS

Condillac's conception of analysis plainly owes much, either directly or indirectly, to Descartes, but in one important respect he goes beyond Descartes. This marked feature of originality is to be found in the rôle which he assigns to language. Analysis is possible only by means of language. Condillac may have opposed himself to Descartes' apriorism of innate ideas and asserted with Locke that all ideas have their origin in sensation, but he substituted for Descartes' apriorism another of his own, that of an innate language. This language is an essential condition of the possibility of there being any beginning for human knowledge. "It is necessary that the elements of some language or other, prepared in advance, should precede our ideas; for without signs of some kind it would be impossible for us to analyse our thoughts in such a way as to have any distinct awareness of what we think."[8] The innate language consists of those bodily gestures and attitudes which are the natural expressions of feeling. Condillac calls it "the language of action." Since its nature is determined by the conformation of our bodily organs, members of the same species, sharing similar organs, will express their feelings in the same way and will consequently be able to understand one another whereas members of different species will not.

Today it would seem to be generally acknowledged that all attempts to find the origins of language in the natural expression of feeling have been failures, and that there is an impassable gulf separating these symptoms of feeling from the artificial symbols for our conceptions of things, the cry of pain or joy from propositional discourse about matters of fact. While Condillac suggests in his opening desscription of the nature of language that discourse originates in these overt bodily symptoms, it is not long before he attributes more to the primitive language of action than a mere giving vent to feeling. Thus he writes, "In him who is not yet acquainted with the natural signs

[7]*La Logique*, I, iv, *Œuvres*, II, 381b f.
[8]*Ibid.*, II, ii, *Œuvres*, II, 396b.

determined by the conformation of the organs, the action makes a very composite picture; for it indicates the object which affects him, and at the same time it expresses both the judgment he makes and the feelings he experiences."[9] This implies that besides symptoms of feeling, the most primitive type of expression contains both the symbolization of objects and the formulation of judgments.[10]

But though analysis, or making our ideas clear to ourselves and others, is the function of a deliberately developed language, it is not the function of language considered in its origin. Language begins to become analytic only when men feel the necessity of communicating with one another. Nor do they form any intention of making themselves understood by others until after they have observed that they are in fact understood by them. Primitive expression is extremely confused because it expresses everything that is thought and felt simultaneously. It is only by decomposing the expressive movements of others, distinguishing one after the other the constituent elements of complex actions, that men come to understand each other's different feelings and thoughts.

Each will notice then sooner or later that he never understands others better than when he has decomposed their actions, and as a result he will come to see that if he is to make himself understood by others he will have to decompose his own actions. Then little by little he will form the habit of repeating successively the movements which nature would cause him to perform all at once, and the language of action will become naturally for him an analytic method. . . .

In decomposing his action he decomposes his thought for himself as well as for others; he analyses it and makes himself understood because he understands himself.

Just as the total action is a picture of his whole thought, so equally the partial actions are pictures of the ideas which make up the parts of his thought. Thus, if again he decomposes his partial actions, he will at the same time be decomposing the partial ideas of which his actions are the signs, and he will continually be forming new distinct ideas.[11]

---

[9] *Cours d'études, Grammaire,* I, ii, Œuvres, I, 430a.

[10] In the *Essai sur l'origine des connoissances humaines* (1746), when discussing the dance as a form of the language of action among the ancients, Condillac remarks that "the dance naturally divided itself into two subordinate arts. The one . . . was the *dance of gestures*, which was reserved for the communication of men's thoughts to one another. The other was primarily the *dance of steps*, which was used to express certain states of the soul, particularly joy; they used it on occasions of rejoicing, and its principal object was pleasure." II, i, 1, Œuvres, I, 63a. The distinction is made here between discourse and the expression of feelings, though both take place by movements of the body.

[11] *La Logique,* II, ii, Œuvres, II, 397b.

The signs belonging to primitive language are not deliberately chosen by us, but are determined for us by nature. At the same time, nature has set us on the road to imagining and therefore to choosing signs for ourselves. It is in this way that the use of artificial signs arises. But even if these deliberately chosen signs are artificial, they are not arbitrary, for they are always formed by analogy with the original natural signs. Just as soon as men begin to decompose their thoughts, the language of action begins to become an artificial language. "It becomes every day more artificial, because the more they analyse, the more they feel the need to analyse. To facilitate these analyses they will imagine new signs analogous to the natural signs."[12]

When in *La Langue des calculs* (published posthumously in 1798) Condillac considered the most perfect language, algebra, he attached extreme importance to analogy as the clue to development of a language.

Since algebra is a language made by analogy, analogy which makes language makes methods, or rather the method of discovery is only analogy itself.
It is to analogy that the whole art of reasoning reduces, as also the whole art of speaking, and in this single word, analogy, we see how it is possible to instruct ourselves in the discoveries of others as well as how to make them for ourselves.[13]

Condillac worked out this thesis in a detailed examination of the development of mathematics, beginning with its origin as a language of action, that is to say, as a calculation with the fingers. When, later, numbers and then letters are substituted, analogy with the original operations by the fingers is carefully retained. Condillac's insistence on the development of language by analogy arises from two main concerns. The one is to prevent any rupture between artifice and nature, for nature is our original mentor in matters of method. The language of science may be artificial, but it is not arbitrary, nor merely conventional. Analogy is also important to him as providing support for his theory that the steps of a demonstration are merely successive linguistic transformations of the same proposition. The retention of the identity of the propositions throughout these transformations is the work of analogy.

In algebra Condillac saw "a striking proof that the progress of the sciences depends uniquely on the progress of languages, and that well-

---

[12]*Cours d'études, Grammaire*, I, i, *Œuvres*, I, 431a.
[13]*La Langue des calculs*, Préface, *Œuvres*, II, 420a.

constructed languages alone can give analysis the degree of simplicity and precision of which it is susceptible."[14] The success of a scientific genius is insuperably conditioned by the degree of development in the language of the age to which he belongs. Newton's immense achievement was made possible by the devices of language and the consequent methods of calculation made available to him by his predecessors. Where words are lacking a science will encounter the same obstacles as geometry did before the invention of algebra, and Newton's work would have suffered similar limitations if he had been born in an earlier age.[15]

Different sciences are not by the nature of their subject-matters susceptible to different degrees of precision. There are only different degrees of success in the construction of their languages—"all would have the same precision if they were always spoken with well-made languages. . . . All sciences would be exact if we knew how to speak the language of each."[16] It was Condillac's unachieved ambition, as expressed to his friends, to construct the languages for metaphysics and the moral sciences, and to rescue them from the chaos into which the abuses of language had plunged them. His editors of 1798 wrote "The unintelligible jargons which they too often speak would have been converted into as many beautiful languages, which everyone would easily have learned since everyone would have understood them; and in these languages the ideas which appeared to be the most inaccessible to the human mind would have been seen issuing of themselves and effortlessly from the commonest notions."[17]

Although Condillac conceived of one universal method for all the sciences, he did not propose, like Leibniz, a single language for them all. Each science has its own language. Algebra is but one of these languages, that one which is appropriate to the subject-matter of mathematics. Other subjects would have their equally appropriate languages. While the symbolization employed in mathematics represents for Condillac a wonderful simplification of the language of words, he never suggests, as Leibniz did, that metaphysics or the moral sciences would be able to find a corresponding kind of simplification, or that they would ever depart from calculation with words. There is one method for the sciences, but a plurality of languages, and

---

[14]*La Logique*, II, vii, *Œuvres*, II, 409b.
[15]*Essai*, II, i, 15, *Œuvres*, I, 99b f.
[16]*La Logique*, II, vii, *Œuvres*, II, 409a.
[17]*Œuvres*, II, 529 n.

through their differences of language the sciences remain distinct from one another.

Condillac extends unity of method not only to all the sciences, but also to all arts employing language, whether the language involved consists of bodily movement or of sounds. Thus the arts of the dance, the theatre, music, oratory, poetry, and every branch of literature, are all produced by this one method. *"To invent, it is said, is to find something new by the force of one's imagination.* This definition is thoroughly bad."[18] Invention is not the product of imagination, nor is genius the product of inspiration.

A geometer will tell you that Newton must have had as much imagination as Corneille, since he had as much genius, but he fails to see that Corneille himself had genius only because he analysed as well as Newton. Analysis makes poets just as it makes mathematicians, and although it causes them to speak different languages, it is always the same method. In short, given the subject of a drama, to find the plan, the characters, their speech, means just so many problems to be solved, and every problem is solved by analysis.[19]

The artist, then, is not a creator. His activity consists in finding something which is already there to be found. There is nothing in artistic invention to correspond to the divine creation *ex nihilo.* The artist's invention is only an analysis through the device of language of something given to him. By means of expression in language he becomes the first to see what we as the result of his work are then also able to see.

THE IDENTITY OF METHOD IN ALL THE SCIENCES

A proposition taken by itself is an analysis—the analysis of a complex idea. Such an idea will remain vague and confused until a proposition is used to articulate its components. Consequently, a true proposition asserts only an identity. Since it states what is contained in the complex idea, "it is limited to saying that the same is the same."[20] All true propositions are either self-evident or provable. Those are self-evident in which the identity of the terms is immediately seen.

[18]*La Langue des calculs,* II, i, Œuvres, II, 470a.
[19]*Ibid.,* 470b.
[20]*Cours d'études, De l'art de penser,* I, x, Œuvres, I, 748a.

But one proposition is said to be the consequence of another when, by a comparison of their terms, they are seen to assert the same thing. Hence a proof is only a succession of propositions all asserting the same thing.

To demonstrate is then to translate an evident proposition, to make it take different forms until it becomes the proposition we wish to prove. It is to change the terms of the proposition, and to arrive, by a succession of identical propositions, at a conclusion identical with the proposition from which it is immediately drawn. It is necessary that the identity which is not perceived as we pass over the intermediary propositions should be evident by the mere inspection of terms, as we go directly from one proposition to another.[21]

It is this theory of the proposition which determines that there can be only one method for all the sciences, and Condillac undertakes to show how this method is exactly the same whether we are solving a problem in algebra or in metaphysics.

He first proposes an elementary mathematical problem. We are to suppose that I have a number of tokens in my hands, and that if I transfer one of these from the right hand to the left, I shall have an equal number in each hand, while if I transfer one from the left to the right, I shall have twice as many in the right as in the left. How many tokens have I in each hand? The solution to the problem involves two operations of reasoning. The first consists in what Condillac calls "reasoning on the conditions of the problem"; the second, in the actual finding of the solution. Both operations consist in a series of translations.

In the first operation the initial statement of the problem is transformed through a series of successively simpler expressions until it finally attains its simplest possible expression. For the case under consideration this simplest expression in its verbal form would be

The right, minus one, is equal to the left plus one;
The right, plus one, is equal to two lefts minus two.
A further simplification of expression can be introduced, however, by the substitution of symbols for words, x and y being made to stand for the unknown quantities in the right and left hands respectively, $+$ being made to stand for "plus," $-$ for "minus," and $=$ for "is equal to." We then get the two equations

[21]*Ibid.*, *De l'art de raisonner*, I, i, *Œuvres*, I, 623a.

$$x - 1 = y + 1$$
$$x + 1 = 2y - 2$$

In this operation we have begun with a statement of the conditions of the problem, which contains within it both the known and the unknown factors. It is an absolute requirement that this statement should contain within it *all* the known factors necessary to the solution of the problem; otherwise the problem will be insoluble. The initial mode of stating the conditions does not, however, prepare us for the solution of the problem. Preparation consists in the reduction of the question to its simplest possible expression through a succession of translations. This reduction performs two important functions. In the first place, it clearly determines whether all the knowns necessary for the solution are actually contained in the statement of the conditions. And in the second place, it gives to the statement of the conditions the explicit form of a statement of identity, i.e., formulates it as an equation—or, as in the example provided where there are two unknowns, the reduction results in two equations. This concludes the reasoning on the conditions and the problem is now ready for solution.

The second operation of reasoning, like its predecessor, consists in a series of translations in which we go from identical proposition to identical proposition, or from equation to equation, until we have the unknowns equated with the knowns,

$$x = 7; y = 5.$$

The problem is now solved, and we know that I had seven tokens in the right hand, and five in the left.

Mathematics reasons with equations, but so do all the sciences, for all propositions are equations, and consequently there is no difference in the way in which we reason in the different sciences. In mathematics, however, the statement of a problem usually formulates explicitly the conditions of the problem, whereas in the other sciences this is unusual. If, for example, we ask the metaphysical question, What is the origin and generation of the faculties of the human understanding? the conditions of the question have still to be found.

As we have seen, in every question capable of being answered the conditions must be present in the statement of the question, and the knowns necessary for finding the unknowns must be included in those conditions. In a question capable of solution, if the conditions are not explicitly stated, then they must at least be implicitly contained in the form of the question, otherwise we should never be able to find them. It is true that they are not always equally recognizable in the original form in which questions are asked. To discover them means to translate the initial statement into its simplest terms and, when this is done, the conditions which were contained implicitly in the question will be rendered explicit.

"What is the origin and generation of the faculties of the human understanding?" To get the conditions of the problem, this should be translated into the question, "What is the origin and generation of the faculties by which man, capable of sensations, conceives things by forming ideas of them?" We then see that attention, comparison, judgment, reflection, imagination, and reason are, along with sensation, the known factors in the problem, while the *origin* and the *generation* are the two unknowns. The conditions are now clearly apparent. The first part of the problem is to determine which of the knowns is the origin, i.e., the principle or beginning, of all the other knowns. Very little observation tells one that it is the faculty of sensation which is involved in all the other faculties. Sensation, therefore, is the *origin*. What still remains to be done is to show how sensation becomes successively attention, comparison, judgment, etc. We shall then know their *generation*. And this is what Condillac claims to have done in the first part of his *Logique* (1780).

. . . just as the equations $x - 1 = y + 1$ and $x + 1 = 2y - 2$ undergo different transformations to become $y = 5$ and $x = 7$, so sensation in the same way undergoes different transformations to become the understanding.

The device of reasoning is therefore the same in all the sciences. Just as in mathematics the question is established by translating it into algebra, so in the other sciences it . . . is established by translating it into the simplest expression; and when the question is established, the reasoning which solves it is itself still only a series of translations in which a proposition translating the one preceding it is in turn translated by the one which follows it. In this way evidence passes with identity from the statement of the question to the conclusion of the reasoning.[22]

Since, from this account, method appears to operate wholly within the realm of language, something should be said about the rôle which

[22]*La Logique*, II, viii, *Œuvres*, II, 411a.

Condillac assigns to experience. To begin with, experience supplies the materials on which we reason. But the clear and distinct perception of any matter of fact is a judgment and therefore it is an analysis. Analysis is, however, possible only by means of language. Hence the perception of a matter of fact is identical with the linguistic affirmation of that fact. Thereafter, any reasoning from what is perceived will consist only in the linguistic transformaton of the original proposition in which the perception is expressed. But Condillac also maintains that in an empirical science like physics "the evidence of reason" (i.e., the perception of the identity of two verbally different propositions) and "the evidence of fact" (i.e., observation) should always go together. When we draw the logical consequences of an observation, the evidence of reason needs to be confirmed by new observations; in other words, our transformations of the original proposition expressing an observation should be verified by new observations. Thus, in its second rôle, observation functions as a test of the correctness of our translations.

THE UNITY OF A SCIENCE CONSIDERED AS A SYSTEM

Since the sequence of propositions in a science is only a sequence of translations of an original assertion of an identity, Condillac raises the question whether the sciences are not then simply collections of frivolous propositions. He replies that if there were any thinking being who was not under the necessity of acquiring his knowledge but knew everything already, he would never formulate anything in a proposition. God alone is such a being. For him every truth has the same immediate evidence of identity that $2 + 2 = 4$ has for us. God sees all truths in a single truth, and doubtless all the sciences of which we are so proud appear frivolous to him.

But although every true proposition is identical, it will not necessarily seem to be so when we first consider it. If we were to see the identity immediately, then, of course, it would be frivolous. But if the identity requires to be established, then it can be instructive, and that for two reasons. In the first place, the human mind is capable only of acquiring successively the partial ideas which go to make up a complex idea, and, secondly, our minds are unable to embrace simultaneously in a distinct manner all the partial ideas contained in a complex idea. From the nature of a proposition and of the relations of propositions to one another in a demonstration it follows that any science considered as a logical system is simply the articulation of a single complex idea. "A complete system can only be one and the same idea."[23] Thus, for example, in metaphysics, we see how sensation becomes in turn each of the faculties, and each of the kinds of ideas. It contains a series of propositions which are instructive when taken in relation to us, but which when taken in themselves are identical. The whole science of the origin of human knowledge can be abridged in the proposition *sensations are sensations.* "If in all sciences we were equally able to follow the generation of ideas . . . we should see one truth giving birth to all the rest, and we should find the abridged expression of all our knowledge in this identical proposition, *the same is the same.*"[24]

A detailed working out of this conception of the nature of a science as a system is provided for his royal pupil in several examples. After considering some taken from geometry he points out that he has begun with the definition of the word "to measure," and that this definition is found in all the following propositions, the only thing which changes being the terms in which it is expressed. Indeed all mathematical truths are but different expressions of this first definition. "Thus mathematics is an immense science contained in the idea of a single word."[25]

At the conclusion of his examination of "Newton's system," he points out that a little reflection on the balance, the lever, the wheel, pulleys, the inclined plane, and the pendulum shows that these machines and others more complicated are reducible to a single one, the balance or lever. Their identity is evident. They take different forms for producing different effects, but in principle all are the same machine. The universe is only a great balance.

Now as all machines, from the simplest to the most complicated, are only one and the same machine which assumes different forms in order to produce different effects, in the same way the properties we discover in a series of machines, each more complicated than the previous, reduce to one first property which in being transformed is at once one and multiple.

[23]*Cours d'études, De l'art de penser,* I, x, *Œuvres,* I, 749b.
[24]*Ibid.*
[25]*Cours d'études, De l'art de raisonner,* I, ii, *Œuvres,* I, 627b.

For if there is in the end only one machine, there is in the end only one property. To be convinced you need only to consider that we have ascended from knowledge to knowledge only because we have passed from identical propositions to identical propositions. Now, if we were able to discover all possible truths and to be assured of them in an evident manner, we should produce a succession of identical propositions equivalent to the succession of truths, and as a result we should see all truths reduced to a single one.[26]

Condillac has no doubt that there is one single principle which would explain all observable phenomena, although this principle has not yet been discovered. He combats the claim of the bolder Newtonians that attraction is that principle.[27] Attraction is a phenomenon which explains many others, but it is far from explaining all. It is a principle which presupposes, or appears to presuppose, a principle even more general than itself, and Condillac sees no reason for supposing that with Newton we have reached the limits of knowledge in physics.

Although he knew Leibniz's metaphysical theory of the monad and was one of its severest critics, Condillac probably had little, if any, acquaintance with his logical theory. It is, therefore, of passing interest to note how he shares with Leibniz the same view of the nature of the proposition and of the nature of proof. For both of them every true proposition is either a self-evidently identical proposition or reducible to one. Condillac differs, however, in maintaining that the whole of a science is reducible to a single self-evidently identical proposition. This conception of what makes a science *one* has no counterpart in Leibniz.

THE SYSTEMATIC UNITY OF ALL THE SCIENCES TAKEN TOGETHER

Not only is the individual science a system, but there is a system of all the sciences taken together. It is impossible, however, that the second system should be the same sort of thing as the first, for if it were it would require that we should be able to reach that one truth

---

[26]*Ibid.*, III, xi, *Œuvres*, I, 676b.

[27]Condillac remarks on the great ingenuity with which they have sought to use it to explain such phenomena as solidity, fluidity, hardness, softness, elasticity, dissolution, fermentation, boiling, etc., in the absence of any empirical verification. *Ibid.*, IV, i, *Œuvres*, I, 678 ff.

which includes within it all truths. Condillac believes there is such a truth but that it is known by God alone and its possession is equivalent to omniscience. Thus, although Condillac suggested that the one fundamental truth of physics might in due course be attained by us, he never entertained the possibility of our attaining the fundamental truth which would embrace the truths of all the different sciences. Moreover, as we have seen, a single science consists in the construction of a single language, but there is no proposal such as we find in Leibniz that we might construct the single language of a universal science, containing within it all the particular sciences. The languages of mathematics and metaphysics, for example, remain irreducibly different for Condillac.

It is, then, in a quite different way that Condillac conceives of all the sciences as forming a unified whole or system. His model for this conception is the living organism, with its integrated system of needs and its means of satisfying them. Although Condillac purported to be a follower of Locke's sensationalist theory of knowledge, he retained certain fundamental features of Descartes' and Malebranche's conceptions of sense knowledge. According to Descartes and Malebranche scientific knowledge, which is knowledge of things as they are in themselves, is possible through the intellect alone, even if, as in the case of mathematics and physics, but not metaphysics, the intellect can be aided by the imagination and the senses. In contrast to the scientific knowledge of things as they are in themselves, there is for them both a type of knowledge which is essential for the conduct and self-maintenance of man as a composite being, who is made up of a mind and a body, and who interacts with a physical environment. The foundation of this practical knowledge resides in the senses, whose function is not to inform us of the nature of things as they are in themselves, but of their existence and how they affect us. Through intellect alone we know the essential nature of things; through the senses we know things in relation to us and to our needs.

Condillac's departure from Descartes and Malebranche lies in his assertion that there is no knowledge of things as they are in themselves, that all knowledge is through the senses, and that, therefore, knowledge has for its objects only things in so far as they are related to us and our needs. Thus all knowledge is practical in nature. There is no science of the kind envisioned by Descartes and Malebranche.

Science is based on sensation and therefore its function is to direct the action towards self-maintenance of an organism situated in an environment.

Our faculties have been bestowed on us to show us how to avoid the harmful and to seek what can be useful to us. For this there is no need for us to have knowledge of the essences of things.

The author of our nature does not require it . . . he wills only that we judge of the relations which things have to us and to one another, when the knowledge of these relations can be of some use.

We have a means of judging of these relations, and it is unique; it is to observe the sensations which objects produce in us. Just so far as our sensations can be extended, the sphere of our knowledge can be extended; beyond that all discovery is forbidden us.[28]

It is this intrinsic relativity of all knowledge to our nature which gives to the sciences a systematic unity. Condillac conceives of the human organism and its environment as interacting with a perfect, divinely-contrived co-ordination. Human needs and the means of their satisfaction are a function of this relation, and it is experience, or nature, through pleasures and pains, which very promptly instructs us as to what is necessary for our self-maintenance. Our needs, like the different organs to which they are related, comprise an integrated system, and so also do the teachings of nature. The one system will correspond exactly to the other. "I see in the sphere of my cognitions a system corresponding to that which the author of my nature has followed in making me; and that is not surprising; for, given my needs and my faculties, my inquiries and my cognitions will be given also. Everything is connected in the two systems in the same way."[29] In pursuing scientific inquiry it is important to study this system of needs and faculties, for since all our knowledge is concerned, not with things as they are in themselves, but only as they are in relation to us, it must form a system corresponding to the organization of our nature. Answering to each need there will be the idea of the thing appropriate to its satisfaction, and the analysis of this idea will in turn give rise to a whole chain of ideas, or what is the same thing, a succession of formulations in language of that one idea. Accordingly the total system of knowledge will consist in a series of fundamental ideas, answering to fundamental needs and interrelated in the same

[28]*La Logique*, II, i, *Œuvres*, II, 394a.
[29]*Ibid.*, 393b.

way as these needs, and then, issuing from each of these fundamental ideas, there will be the series of their constituent ideas as elicited by analysis. These constituent ideas, however, will not form isolated series but will be interwoven with one another as they ramify, "for the same objects and the same ideas are often related to different needs."[30] Since any individual science is for Condillac the analysis or progressive articulation of a fundamental idea, there will be a plurality of sciences as there is a plurality of needs, but like the needs they will be co-ordinated in an organic whole. Although Condillac describes the sciences as logically interrelated in this system, the description is incompatible with his definition of a logical system, for the way in which the differentiated functions of the parts of an organism are united in the common service of the well-being or preservation of the organism is quite distinct from the way in which the different terms of a proposition are united in an equation. It is the latter kind of union alone which is the basis of a logical system.

In spite of the fact that no philosopher has ever presented a theory providing a simpler and easier basis for identifying a science than did Condillac—all that was required of him was to name the fundamental idea of which any given science is the analysis—yet so concerned was he for the unity of knowledge that he expressed the strongest disapproval of any attempt to classify or define the different sciences. He despised theory about their divisions as characteristic of scholasticism, and he found the eighteenth century, in spite of the great scientific advances it had made, still dominated by this scholastic passion for distinctions and the separation of intellectual disciplines. "We are more scholastic than we think."[31] His royal pupil is reminded that in Greece a scholar cultivated all the arts and sciences together and his mind was thereby strengthened by all the helps which they mutually supply to one another. The superiority of the Greeks to the Romans was owing to this universality in their approach to knowledge. "Nature shows us by a thousand examples that there are things which must not be studied separately. Indeed, a grammarian will never be anything but mediocre or bad if he is only a grammarian. It is the same with a rhetorician, a logician, etc. We shall ourselves, therefore, be badly instructed in these arts if we study them separately."[32] To

[30]*Cours d'études, De l'art de penser*, I, iv, *Œuvres*, I, 726b.
[31]*Cours d'études, Histoire moderne*, VIII, vii, *Œuvres*, II, 155b.
[32]*Ibid.*, 154a.

study them in this way is to separate things which by their nature are intended to throw light on one another. But, one might ask, if the sciences are not pursued separately, will they not all end by becoming confused? Condillac agrees that ultimately, when our intellectual possessions become too many to be embraced together, specialization is necessary. Nevertheless, in their beginnings the different branches of knowledge must be taken together, as they were by the Greeks. Unfortunately, instead of trying to find out how the Greeks arrived at their knowledge, we have concerned ourselves only with the results of their inquiries. Instead of trying to recover a course of education in which all our studies are blended, we have slavishly followed the order and divisions of the scholastics, and in the end have succeeded in producing even more divisions than they, ontologies, psychologies, cosmologies, and so one. Despite his admiration for Bacon, Condillac considered him particularly guilty of this excessive multiplication of divisions in the sciences.[33]

In conclusion it may be remarked that on its logical side Condillac's theory of unity of method fractures knowledge into a set of sciences each of which is reducible to a single member of a set of logically discrete and independent identical propositions. But on its utilitarian side, in which inquiry is directed to the means of satisfying human needs, his method integrates all knowledge into one whole of mutually interrelated parts.

[33]Ibid., XX, xii, Œuvres, II, 230b.

VI. DIDEROT AND D'ALEMBERT: THE ASSAULT ON "L'ESPRIT DE SYSTÈME"

THE PROBLEM OF FINDING A SYSTEM OF UNIFICATION

Diderot was to express again Leibniz's alarm over the immense and haphazard proliferation of learning continuously being deposited in the world of books. "As the centuries pass the mass of books will constantly increase, and one can foresee a time when it will be almost as difficult to learn anything in a library as it would be in going directly to the universe itself."[1] Whenever this stage is reached men will have to devote themselves to the work of rediscovering what has already been discovered without knowing that it has already been done—"for if we are now ignorant of a part of what is contained in so many volumes published in all sorts of languages, we shall know even less of what is contained in these volumes when a hundred, a thousand, times as many more have been added to them."[2] The reduction of this accumulated knowledge to manageable proportions by making extracts of the important parts and giving them a systematic arrangement in an encyclopaedia was the task to which Diderot and his colleagues set themselves. "We have undertaken now for the good of the world of learning and in the interests of the human race a work to which our grandsons would be compelled to devote themselves, but under much less favourable circumstances, when the superabundance of

[1]"Encyclopédie," *Encyclopédie ou dictionnaire raisonné des sciences, des arts et des métiers, Œuvres complètes de Diderot*, ed. J. Assézat (Paris, 1875-7), XIV, 475 f.
[2]*Ibid.*

books will have rendered its execution most difficult."[3] For Diderot, as for Leibniz, the supreme value of the encyclopaedia as an instrument for the preservation of civilization would at once be seen should some great catastrophe occur which would "suspend the progress of the sciences, interrupt the work of the crafts, plunge a portion of our hemisphere once more into darkness. What gratitude would not be bestowed by the generation which came after the times of trouble upon the men who had anticipated them from afar and forestalled the havoc by rendering secure the knowledge of past ages."[4]

In his *Prospectus* for the *Encyclopédie* Diderot presented what he called "the system of human knowledge"[5]—a general organization of the arts and sciences based on Bacon's, though deviating from it in several significant respects. The *Prospectus* (1750) and d'Alembert's *Discours préliminaire* (1751) show an almost exclusive concern with the idea of an encyclopaedia as the unification of knowledge in a system, and with determining the nature of this system. However, in the article "Encyclopédie" which appeared in volume V (1755) of the *Encyclopédie*—that is, after the project had been under way for some time—Diderot was much more explicitly concerned with another kind of unification, one which proceeded so empirically as to be, if not incompatible with system-construction, at least not specifically related to it. This latter mode of unification was much more in keeping with the extreme distrust of *l'esprit de système* which, as we shall see, animated both him and d'Alembert.

But there were practical difficulties which were to render the great idea of a system of all knowledge virtually impossible. In the article "Encyclopédie," Diderot writes:

When we come to consider the immense subject-matter of an *Encyclopaedia*, the one thing which is clearly seen is that it cannot be the work of a single man. How could one man, in the short period of a lifetime, succeed in knowing and explaining in detail the universal system of nature and of art, when the large and learned society of the academicians of *la Crusca* took forty years to make its dictionary, and our own French academicians worked sixty years on their Dictionary before publishing the first edition? . . .

A single man, you may say, can be master of all the knowledge which exists; he can dispose as he likes all the riches that other men have accumulated. I cannot agree with this supposition; I do not believe that

[3]*Ibid.*   [4]*Ibid.*, 428.   [5]*Œuvres*, XIII, 145.

one man is capable of knowing all that can be known, of making use of all of it that exists, of seeing all that can be seen, of understanding all that is intelligible.[6]

How then would it ever be possible to comprehend in one synoptic view the immense wealth which lies beyond the power of any one man to master? There was, as well, the fact that the enormous number of contributors, though they formed what Diderot called "a society," were not bound together by any unity of doctrine. Their contributions abounded in mutual contradictions, which he freely acknowledged. There were to be no meetings and no transactions among the members of this society. Indeed, it was an essential part of Diderot's plan that each should work independently of the others, with no common tie but that of concern for the general interest of the human race and by a feeling of mutual goodwill. It was this common concern alone which made them into a society. The result of the plan, whereby each contributor was to be occupied solely with his own special subject and make no encroachments upon the territory of another, was that the role of the editor, as Diderot said, was reduced to very little.[7] And d'Alembert summed it up: "In a word, each of our colleagues has made a Dictionary of the part with which he was charged, and we have put all these Dictionaries together."[8]

This quite evidently is not a procedure by which systems are constructed. If, as Diderot says, "The word *Encyclopaedia* signifies the unification of the sciences,"[9] then the unity which he sought to achieve in his own project was something other than that of unity of system, or the distribution of elements in a great architectonic scheme, such as had been envisaged by Leibniz. Nevertheless, in his *Prospectus* and in d'Alembert's *Discours préliminaire*, when there is discussion of the nature of an encyclopaedia, the problem of unification presents itself to both authors without any question as one of finding some mode of co-ordinating all the diverse kinds of human knowledge into a system, and system construction still remains a subject of discussion in Diderot's article, "Encyclopédie." D'Alembert points out in the *Discours* that an encyclopaedia is not just a dictionary. The function of a *dictionary* of the arts and sciences would be to present the general

[6]"Encyclopédie," *Œuvres*, XIV, 416.
[7]*Prospectus*, *Œuvres*, 135.
[8]*Discours préliminaire de l'Encyclopédie*, ed. F. Picavet (Paris, 1894), 133.
[9]"Encyclopédie," *Œuvres*, XIV, 414.

principles which constitute the foundation of each, together with the most essential part of its contents. The function of an *encyclopaedia*, on the other hand, would be to exhibit the unity of human knowledge. Their own *Encyclopédie, ou dictionnaire raisonné des sciences, des arts et des métiers*, to give it its full title, was designed to combine both these functions in a single work.[10]

Having stated the task of unification, d'Alembert, as we have said, immediately interprets it as a problem of finding a "system," and notes its extreme difficulty. It is easy enough to see that the sciences are interconnected. "But if it is often difficult to reduce each particular science or art to a small number of rules or general notions, it is not less so to comprehend within a single system the infinitely varied branches of human science."[11] Both Diderot and d'Alembert emphatically repudiate the notion that there is any such thing as one objectively valid system of *human* knowledge, though there may be such a thing for divine knowledge. There are numberless possible systems, and the construction of any one of them involves a high degree of arbitrariness. For example, says Diderot, a system can be constructed by classifying the different kinds of knowledge according to the faculties of the mind or, on the other hand, according to the kinds of being which are the objects of knowledge. Another classification could be based on the distinction between the science of things and the science of signs; still another on that between the science of concretes, and the science of abstracts. Or one could divide knowledge according to the two most general kinds of cause, namely, art and nature. Equally well the distinction between the physical and the moral could be taken as the basis for a system, or that between the existent and the possible, or the material and spiritual, or the real and the intelligible, or what has its origin in sense and what has its origin in reason, or the natural and the revealed.

It is then impossible to banish the arbitrary from this first great distribution. The universe presents us only with particular beings, infinite in number, and almost without any fixed or determinate division; there is no one of them which can be called either first or last; all are connected and follow one another with imperceptible gradations; and if, across this uniform and vast extension of objects, there appear some which, like the tops of rocks, seem to rise above the surface and to dominate it, they owe this privilege only to particular systems, to vague conventions, to

---

[10]*Discours*, 12 f.   [11]*Ibid.*

certain irrelevant circumstance, and not to the physical arrangement of existing things and to the intention of nature. . . . Both the real world and the intelligible world of ideas can be represented from an infinite number of points of view, and the possible systems of human knowledge are as numerous as these points of view. The only one containing nothing arbitrary would be, as we have said in our *Prospectus*, the system which existed from all eternity in the will of God; and this system in which a descent is made from this first eternal Being to all those beings which in the passage of time have emanated from his breast would resemble the astronomical hypothesis in which the philosopher thinks of himself as transported to the centre of the sun in order to calculate the observable motions of the celestial bodies surrounding him. Such a scheme has simplicity but it could be criticized for a defect which would be serious in a work composed by philosophers and addressed to all men in all time to come: the defect of being too closely tied to our theology, a sublime science, which is useful no doubt for the knowledge the Christian receives from it, but even more useful for the sacrifices it demands and the rewards it promises him.

As for this general system from which anything arbitrary is excluded, and which we shall never possess, there would perhaps be no great advantage in having it; for what difference would there be between reading a book in which all the actuating principles of the universe were explained in detail and the study of the actual universe itself? Almost none. We should never be capable of understanding more than a certain portion of this great book; and just so far as the slightest impatience and curiosity which overcome us and so commonly interrupt the course of our observations threw our reading into disorder, our knowledge would become as fragmentary as it is now. Losing the chain of inductions and no longer perceiving the connections with what goes before and after, we should soon have the same gaps and the same uncertainties. At present we occupy ourselves in filling these gaps by contemplating nature; then we would be occupying ourselves in filling them by meditating on an immense volume. And this volume, since it would not seem more perfect to us than the universe, would not be any less exposed to the rashness of our doubts and our objections.[12]

In making his own choice of a primary principle of division for organizing knowledge Diderot puts man in the place which God would occupy in "the absolutely perfect plan"—a plan which, as he said, would be of no use to us in any case. For,

. . . if man, this thinking and contemplating being, were banished from the face of the earth, the affecting and sublime spectacle of nature would no longer be anything but a sad and silent scene; the universe would be

[12]"Encyclopédie," *Œuvres*, XIV, 450–2.

stilled, silence and night would take over. It is the presence of man which gives significance to the existence of things. Can any better plan for the history of these things be proposed than to submit to this consideration? Why not introduce man into our project, in the same way as he is placed in the universe? Why not make him a common centre of it? . . . That is why we have sought in man's principal faculties the general division under which we have brought our work. . . . Man is the unique point from which we must start and to which everything must be led back, if we wish to please, to inspire with interest, to evoke any feeling, even in the most arid considerations and the driest details. If you take away my existence and the happiness of my fellow creatures, what matters the rest of nature?[13]

Thus, relating all knowledge to the human faculties, Diderot takes over Bacon's primary division according to memory, reason, and imagination.[14]

D'Alembert describes the function of encyclopaedic organization as that of raising the philosopher to a sufficient height above the great labyrinth of human knowledge that he can see it all in a single comprehensive view. An encyclopaedic system is therefore a kind of mapmaking. As in the construction of maps of the whole globe so also with the encyclopaedia it is necessary to adopt some particular position from which to look at the globe and decide on some scheme of projection. Systems of knowledge can be as numerous as systems of projection; and each can claim to have certain advantages which are exclusive to itself. Whichever one we choose, however, we can never be sure that it is the best one; that is to say, the one which would indicate the greatest number of connections between the different sciences. One and the same thing can always be classified in numberless ways, and there is necessarily, therefore, an element of the arbitrary in any general system of division. The division which seemed best to Diderot and d'Alembert was the one which most effectively combined the encyclopaedic order with the order in which our different kinds of knowledge have their genesis. For this reason they adopted

[13]Ibid., 453.

[14]The order was changed so that reason should come before imagination. This was done on the grounds that the new order corresponds more accurately to the order of development of the mind's powers. The imagination is a creative faculty, but before the mind can create it must be able to reason on what it sees and knows. Furthermore, in the imitation of nature in the fine arts invention is subject to certain rules; it is not the work of genius alone. But rules are the product of reason. Discours, 64.

Bacon's system of classification, though introducing into it certain modifications: "But we are too convinced of the arbitrariness which always governs in such a division to believe that our system is either the only one or the best."[15]

THE CLASSIFICATION OF THE SCIENCES

Since our main concern here is with the Encyclopaedists' conception of the nature of a system of all knowledge, rather than with its actual ramifications and details, we shall look only at that division of knowledge which corresponds to the faculty of reason, namely, philosophy or science, and observe there the principles in accordance with which the particular sciences are ordered in relation to one another. It was in this branch of knowledge that Diderot and d'Alembert claimed to owe least to Bacon.

The system which they adopted was a hierarchical one, in which the different levels are levels of abstraction. It was designed to accommodate "the encyclopaedic order" to "the genealogical order" of knowledge, or the order in which ideas develop. "The natural progress of the human mind," says Diderot, "is to rise from individuals to species, from species to genera, from proximate genera to remote genera, and to form at each level a Science; or at least to add a new branch to each Science already formed. . . ."[16] Thus at the very highest level of abstraction there will be a science which is concerned with the most general properties of beings, whether spiritual or material, such as *"existence, possibility, duration, substance, attribute,* etc." This is ontology, or the science of being in general.

At the next level below ontology we have two sciences distinguished from one another according to the kinds of being with which they are concerned. Since there are two kinds of being, spiritual and corporeal, we accordingly have a most general science of spiritual beings, and a most general science of corporeal beings. These are called respectively pneumatology or particular metaphysics, and general physics or the metaphysics of body. Metaphysics of body will deal with the most general properties common to all bodies, such as extension, motion, impenetrability, etc. Quantity, which is one of these, is separated from the metaphysics of body to become the object of a special science, mathematics. Mathematics in turn has its

[15]*Ibid.*, 62. [16]*Prospectus, Œuvres,* XIII, 149.

levels of abstraction. There is pure mathematics, which deals with quantity in the abstract, and mixed mathematics which deals with quantity as it exists in things, giving rise thereby to such sciences as mechanics, astronomy, optics, acoustics, etc. Then, finally, there is particular physics (as contrasted with general physics or the metaphysics of body), and its divisions will correspond to the divisions of natural history.

Following on pneumatology, or the general science of spiritual beings, there will be the sciences which are concerned with the different kinds of spiritual beings, God, angels and demons, and the human soul—hence natural theology, the doctrine of good and evil spirits, and the science of man. The distribution of the science of man is determined by the faculties of his mind, namely, understanding which must be directed towards truth, and will which must be directed to virtue. Accordingly, on the one hand, there will be logic, with all its subdivisions including, for example, the art of thinking, grammar, and rhetoric, and, on the other hand, ethics, with its subdivisions. There will be general ethics, and then the particular moral sciences: natural jurisprudence, economics, and politics.

In comparing this classification of the sciences with Bacon's, as we are invited to do by Diderot, we can discern several striking differences. In the first place the very purpose of the classification is different. The Encyclopaedists undertake it in order to exhibit the unity of the sciences. This is not a rôle which is ever assigned to it by Bacon, who is concerned rather with making such an inventory of knowledge as will reveal where the gaps and deficiencies are and thereby direct scientific effort towards work which needs to be done. For giving unity to the sciences he looks to *philosophia prima* and to natural philosophy. Turning to the contents of the Encyclopaedists' classification we find the deviation from Bacon's most manifest in their inclusion, first, of ontology or the science of being in general, and, second, of the sciences of spirit. Diderot's conception of ontology or *philosophia prima* is borrowed from Wolff and has nothing in common with Bacon's *philosophia prima*.[17] There is no science of being as such for Bacon. There is only the science of nature. If "being" is enumerated

[17]See "Ontologie" in the *Encyclopédie* for Diderot's high praise of Wolff's ontology. "To him must go the credit for being the first to deal with it in a truly philosophical manner." *Œuvres*, XVI, 167.

by him among the transcendentals which form the subject matter of *philosophia prima*, it is only in so far as it and the other transcendentals are taken "as they have efficacy in nature."

As for metaphysics conceived as the science of spirits, no such thing is possible according to Bacon, for it is closed off by revealed religion from natural inquiry. The only metaphysics in Bacon's classification is that which the Encyclopaedists identify with general physics. As a philosopher Bacon is, as we have seen, a thorough-going materialist. Diderot too is a materialist, and he is indeed claimed by the Marxists as one of "the great materialists,"[18] but his materialism plays no rôle in determining the unity of the sciences in the *Prospectus*, whereas it plays a most decisive rôle in Bacon's conception of their unity. A complete failure to recognize the significance of this materialism in the classification borrowed from Bacon is reflected in d'Alembert's statement, "We do not know why the celebrated author who has served as our guide in this distribution has placed nature before man in his system. On the contrary there would seem to be every reason for placing man at the point of transition from God and spirits to bodies."[19] The answer to d'Alembert's problem is given when Bacon, approaching "the doctrine concerning man," asserts that the knowledge of man—that part of him which is the object of legitimate inquiry—is "but a portion" of the knowledge of nature itself. Natural philosophy was for Bacon the root of all the sciences, including "moral and political philosophy, and the logical sciences." This grounding of the sciences of man in natural science formed no part of the Encyclopaedists' conception of the unity of the sciences.

"L'ESPRIT DE SYSTÈME" AND "LE VÉRITABLE ESPRIT SYSTÉMATIQUE"

In the end, the Encyclopaedists attached very little value to unification by system. So little, in fact, that it has hardly seemed worth while to describe their "system," except for the fact that their continuously expressed lack of conviction about it serves to give rise to a new and different conception of the unity of the sciences. This conception

---

[18]Lenin puts him in this class along with Feuerbach, Marx, and Engels—see his *Materialism and Empirio-Criticism, A Handbook of Marxism*, ed. Emile Burns (London, 1935), 643.
[19]*Discours*, 66.

gradually takes shape as they formulate their devaluation of, and opposition to classification, to the "abstract" sciences as opposed to the knowledge of fact, and to systems, at first to hypothetical systems, but finally and by easy steps to almost any kind of systematization at all.

"We have no wish," says d'Alembert, "to resemble that crowd of naturalists whom a modern philosopher[20] has so rightly censured, and who, while ceaselessly engaged in dividing nature's productions into genera and species, have consumed in this work time which would have been much better used in the study of those productions themselves."[21] Any system for the general division of the sciences is no substitute for the study of the objects of the sciences, and taken for itself alone is merely frivolous. "A single reasoned article on a particular object of science or art contains more substance than all the divisions and subdivisions that can be made with general terms."[22] Of the two functions to be served by their project, that of an encyclopaedia and that of a dictionary, it is the latter, says d'Alembert, which is so much the more important. It is the latter which is of concern to most of their readers and to it the editors have devoted most of their care and effort. This view, that the importance of the elements far outweighs that of their systematic co-ordination, is one of the principal themes of Diderot's De l'Interprétation de la nature (1753).

"Facts, whatever their nature, are the true riches of the philosopher," says Diderot.[23] Philosophers or scientists have been much too preoccupied with the organizing and interrelating of facts, when it is the discovering and assembling of facts which is significant. The "proud architects" of systems have all had their edifices demolished by the passage of time. "Happy the systematic philosopher to whom nature has given, as she formerly did to Epicurus, Lucretius, Aristotle, and Plato, a vivid imagination, great eloquence, the art of presenting his ideas in striking and sublime images! The edifice he has constructed may fall some day, but his statue will remain standing amid the ruins."[24]

The supreme value which Diderot attached to the knowledge of individual facts leads him to make a remarkable attack on mathematics, the greatest achievement of the purely rational and systematic mind. Mathematics is only a game, and the thing with which it concerns

---

[20]Buffon.   [21]Discours, 62 f.
[22]Ibid., 74.
[23]De l'Interprétation de la nature, XX, Œuvres, II, 19.
[24]Ibid., XXI, Œuvres, II, 20.

itself has no real existence in nature. The mathematicians who pour scorn on metaphysics have failed to realize that the whole of their own science is itself nothing but a metaphysics. "It was asked once: What is a metaphysician? A mathematician replied: It is a man who knows nothing. Chemists, physicists, naturalists, and all those who have devoted themselves to the experimental art, while not less excessive in their judgments, seem to me to be on the point of avenging metaphysics, and of applying the same definition to the mathematician. They say: Of what use are all these profound theories of heavenly bodies, all these enormous calculations of rational astronomy, if they do not exempt Bradley or Le Monnier from observing the heavens?"[25] Diderot predicted that within a hundred years there would not be three great mathematicians left in Europe for mathematics would have lost its usefulness.

The possibility of systematizing our knowledge of nature is reduced by Diderot to almost the minimum which is consistent with there being any natural science at all. The powers of the mind are weak, phenomena are infinite, causes are hidden, and the forms of nature are transitory and in continual flux. *De l'Interprétation de la nature* is Diderot's contribution to the assault on *l'esprit de système*,[26] but it goes further. It would seem to leave almost nothing even for *le véritable esprit systématique* which d'Alembert sought to distinguish from the tendency to construct conjectural theories, a tendency which he asserted had received its death blows in Condillac's *Traité des systèmes*.

But how much did d'Alembert himself leave to the true systematic spirit? His conception of it originates in his reflections upon the nature of mathematics as a science, and as the science furnishing the model for all the sciences. These reflections are almost indistinguishable from Condillac's. It may be remembered also that d'Alembert, unlike Diderot, was himself a mathematician. He remarks that in mathematics we seem to know an almost inexhaustible number of truths but, in fact, when we look at this multiplicity of truths philosophically, we find that they are not really different truths, but merely different formulations of the same truth, or, at the most, of a very small number of primitive truths. If we examine a succession of mathematical pro-

[25]*Ibid.*, III, *Œuvres*, II, 10 f.
[26]"Two main obstacles have long held up the progress of philosophy: authority and the systematic spirit." "Philosophie," *Encyclopédie*, *Œuvres*, XVI, 288.

positions deduced from one another, we see that they are nothing but the first proposition successively transformed in a series of translations. There has been no multiplication of truths, but only of modes of articulating one truth.

Like Condillac, d'Alembert carries this conception over from mathematics to physics. Our knowledge of nature appears to be made up of a great many detached truths. We seem, for example, to know a great many different things about electrical bodies, but this multiplicity is not due to the fact that we know so much about them, but to the fact that we know so little. It is to the weakness of our understanding that we owe our "sad advantage" of knowing so many different things, "and one can say that our abundance in this respect is the effect of our very indigence." "The universe, for any one who could embrace it from a single point of view, would be, if one may say so, only one unique fact and one great truth."[27]

The whole aim of science is the reduction of a multiplicity of phenomena to a single phenomenon, which may then be regarded as the principle of the others. The fewer the principles, the greater their extension, and the greater their fruitfulness. "This reduction . . . constitutes the true systematic spirit, which one must be very careful not to confuse with the spirit of system. . . ."[28]

Where the spirit of system differs from the true systematic spirit is not in the type of logical structure it seeks, but in allowing itself to go beyond what is revealed in experience, by speculating and by forming hypotheses. The numerous, but independently known properties of the magnet may, for example, be ultimately explicable by a single general property from which they originate. The search for it is a task worthy of the scientist, but his search must always be undertaken with due regard for the limited powers of the human mind. To form any conjecture as to what it is, is to become a victim of the spirit of system. In his search for unity the scientist must be prepared to have this unity denied to his efforts. "The only resource left to us then in so difficult a quest, however necessary and even agreeable it may be, is to amass as many facts as possible, to dispose them in the most natural order, and to carry them back to a certain number of principal facts of which the others are only consequences."[29]

[27]*Discours*, 39.   [28]*Ibid.*, 30.
[29]*Ibid.*, 31. For the same notions as expressed by Diderot, see *De l'Interprétation de la nature*, XLV, *Œuvres*, II, 42.

It is, then, mathematics which supplies the paradigm for unity of system for d'Alembert. But the paradigm represents an unattainable ideal. All that remains for the scientist is the collecting of facts, classifying them, and eliciting piecemeal such logical connections between them as are capable of being discerned. The "true systematic spirit" does not take the scientist very far on the road to that systematic unity which he is compelled to seek. Human knowledge is condemned to remain confined to a multiplicity of detached and independent truths. Their ultimate unity will forever remain concealed.

THE ALTERNATIVE TO UNITY OF SYSTEM

Such, then, being the attitude of the Encyclopaedists to system in general, it is not to be expected that the notion of system should play any significant rôle in their practical efforts to co-ordinate the various kinds of human knowledge. What can be expected is that they should proceed in the same empirical and piecemeal fashion in dealing with the relations between the different sciences as they would in co-ordinating the facts which fall within the limits of any one of those sciences.

And, in fact, instead of system construction what we have is only the indication of such lines of continuity as happen to offer themselves between what falls within one science and what falls within others. Such lines of continuity are determined by no *a priori* scheme whatever, and taken together they give rise to no pattern, no discernible architectonic, no synoptic view—only an imperfect network. The unity of human knowledge is not a unity in diversity, nor the unity of a whole in relation to its parts, but unity of continuity among parts wherever such continuity may be found. Diderot's conception of this continuity of all knowledge is a mirror of his conception of the continuity of nature. Nature possesses no fixed or determinate divisions. Everything is interconnected by insensible nuances. The same is true of human knowledge. There are no fixed boundaries between the sciences, and one passes by imperceptible gradations from one to another. "All the sciences encroach upon one another. They are continuous branches and stem from the same trunk."[30] "Thus grammar will constantly refer to dialectic, dialectic to metaphysics, metaphysics

[30]"Encyclopédie," *Œuvres*, XIV, 467.

to theology, theology to jurisprudence, jurisprudence to history, history to geography and chronology, chronology to astronomy, astronomy to geometry, geometry to algebra, algebra to arithmetic, etc."[31]

In his *Essai sur les éléments de philosophie*, after dealing separately with the elements of logic, metaphysics, and ethics, d'Alembert sets out to show that knowledge nevertheless ramifies without regard to these divisions or to any logical precedence of one science to another. "Although we have separated these different sciences in order to consider each more particularly, taking account of the nature and the differences of their objects, they are nevertheless more closely united and exercise a greater reciprocal influence than is imagined, and for this reason the most philosophical order one can follow in treating of them well is perhaps less to take them separately than to make them advance together and fit in with one another."[32] Metaphysics is concerned with examining the generation of our ideas.[33] Among these will be our ideas of just and unjust. Thus the primary truths of metaphysics are essentially connected with the primary notions of ethics, and cannot be separated in any philosophical analysis. Logic is the art of comparing ideas, and to know how to compare them we must know how they are generated. Thus, taken from this point of view metaphysics should precede logic. But at the same time the generation of ideas cannot be developed without the use of logic, and so, from this point of view logic should be prior to metaphysics. "It is, then, plainly impossible to treat separately and distinctly any one of these three sciences, logic, metaphysics, and ethics, without supposing several notions already acquired in the two others."[34] There is another very important science which cannot be separated from logic and metaphysics, namely, philosophical grammar, because of the essential rôle performed by language in the art of reasoning and in the analysis of ideas.

Having seen what the Encyclopaedists regarded as the unity of human knowledge, we must now turn to consider the means they adopted to establish this unity. The contributors had been required

[31]*Ibid.*

[32]D'Alembert, *Mélanges de littérature, d'histoire, et de philosophie* (Amsterdam, 1773), IV, 143.

[33]"The generation of ideas belongs to metaphysics. This is one of its principal objects, to which perhaps it should be limited. Almost all other questions it proposes are insoluble or frivolous." *Ibid.*, 45.

[34]*Ibid.*, 144.

by Diderot to work independently of one another upon their assigned subject-matters. The sum total of these contributions, however, comprised only a dictionary. The function of the editor was to transform this dictionary into an *encyclopaedia*; in other words, to exhibit the unity of human knowledge. The instrument for effecting this unification was the *cross-reference*. Diderot described the use of cross-references as "the most important part of our encyclopaedic scheme." He writes,

I distinguish two kinds of cross-reference; the one for things, the other for words. The cross-references to things clarify the subject, indicate its close connections with those subjects which are immediately contiguous to it, as well as its distant connections with others which might be thought to have no bearing on it. They call to mind common notions and analogous principles. They strengthen implications, connect up the branch to the trunk, and give to the whole that unity so favourable to establishing truth and winning its acceptance. But, when necessary, they will also produce a quite contrary effect; they will put notions in opposition to one another and will make principles conflict; they will attack, loosen, and secretly overthrow certain ridiculous opinions which no one would dare openly to affront. If an author is impartial he will always have the double function of confirming and of refuting, of disturbing and of conciliating.

There would be room for great skill and an infinite advantage in this latter type of cross-reference. The entire work would take on an internal force and secret utility, the silent effects of which would necessarily come to be felt in the course of time. Whenever, for example, a national prejudice merited consideration, it would be necessary, in the particular article dealing with it, to expound it respectfully, along with all its trappings of plausibility and enticement; but at the same time to overthrow the edifice of mud and to disperse the vain accumulation of dust by making cross-references to the articles in which solid principles serve as the foundations of directly opposing truths. This method of undeceiving men operates very promptly upon good minds; and it operates infallibly and with no unhappy consequences, secretly and quietly upon all minds. This is the art of tacitly deducing the boldest consequences. If these cross-references of confirmation and of refutation are worked out well beforehand and prepared with skill, they will give to the *Encyclopaedia* the character required of a good dictionary, that of changing common ways of thinking."[35]

The individual contributions to the Dictionary do not have to be made by men possessing common principles and sharing similar doctrines; by means of interrelating the multitude of contributions

[35]"Encyclopédie," *Œuvres*, XIV, 462–3.

through cross-references, truths will be made to reinforce truths and to sift and eliminate what is false.

Besides the cross-references to things there are the cross-references to words. Every science has its own language. But if each science stayed within the confines of its own language, it would be necessary for the sake of clarity for every one of its words to be defined. The consequence would in each case be an exposition requiring so many parentheses and digressions that it would become diffuse, obscure, and utterly tedious. But this could be eliminated by cross-references between the suppositions of different arts and sciences. A word which is merely an accessory in one subject may be the important one in another, and to get its meaning we should be referred to this other subject. With the progressive establishment of community of language among the sciences through the device of cross-references to words, the sciences themselves become more closely linked to one another. Because language is the repository of all that a people know and have discovered, without community of language men's achievements will remain isolated.

VII. KANT:
THE "CONCEPTUS COSMICUS"
OF PHILOSOPHY

THE NECESSITY OF CLEARLY
DISTINGUISHING THE SCIENCES

When considering Leibniz's conception of the unity of the sciences our attention was directed almost entirely to the logical aspects of this unity. This was because the account was restricted to his own explicit references to it, almost all of which are in terms of the demonstrative encyclopaedia and of its instrument, the universal characteristic. But if it is only as logician that Leibniz is explicit, as a metaphysician he is also concerned with the unity of the sciences, and this concern is to be found constantly at work in his attempts to co-ordinate and reconcile such separate spheres of inquiry as metaphysics, physics, mathematics, natural theology, and ethics. Where, as logician, he denied the possibility of any fixed boundaries between the sciences, and attributed all distinctions to names instituted for human convenience, he was compelled, as metaphysician, to demarcate them sharply—for it was only by doing so that he could reconcile the kingdoms of nature and of grace, efficient causes and final causes, the laws which govern substances or souls and the laws which govern material phenomena. Only a separation of the kinds of questions falling within different sciences could remove apparent conflicts which arise from intermingling these questions. Thus in the *Discourse on Metaphysics* (1686), §x, he argues that the conflict between the discredited scholastic physics, with its substantial forms, and the new physics, with its mechanical principles, was not in reality one between two rival theories, each purporting to provide answers to fundamentally the same type of questions. It arose from confusing questions appropriate to different spheres of inquiry, metaphysics and physics. He goes on to discuss the importance of the general principle of determining the limits of any science in order to exclude irrelevant questions. Only when these lines are clearly drawn is it possible to see the connections of the sciences with one another. Berkeley's *De Motu* (1721) is another and brilliant essay on the errors in thinking which result from confusing categories and importing into one science, physics, concepts which are relevant only in another, metaphysics. It ends with the advice: "Allot to each science its province; assign its bounds; accurately distinguish the principles and objects belonging to each. Thus it will be possible to treat them with greater ease and clarity."[1]

Such discussions as these of Leibniz and Berkeley are rare among reflections on the sciences in the seventeenth and eighteenth centuries until we reach Kant, who shows a continual preoccupation in his writings with systematically removing the confusion of concepts belonging to different sciences. He lays it down as a principle that "Every science is a system on its own right; . . . we must . . . set to work architectonically with it as a separate and independent building. We must treat it as a self-subsisting whole, and not as a wing or section of another building—although we may subsequently make a passage to or fro from one part to another."[2] As an example of the violation of this principle Kant cites the introduction of the conception of God into natural science in order to explain purposiveness in nature, and then the use of this purposiveness to prove the existence of God. This mixing of natural science and theology means that both sciences are "deprived of all intrinsic substantiality. This deceptive crossing and re-crossing from one side to the other involves both in uncertainty, because their boundaries are thus allowed to overlap."[3] Or again, there is the careless kind of thinking by which logic gets mixed up with psychology, or is allowed to borrow some of its principles from that science; or ethics gets confused with anthropology; or the practical principles of morality are thought to be practical in the same sense that the technical skills based on natural science are practical; or the method of philosophy is thought to be the same as that of mathematics;

---

[1]*De Motu*, § 72, *The Works of George Berkeley*, ed. A. A. Luce and T. E. Jessop (Edinburgh, 1948–57), IV, 52.
[2]*Critique of Teleological Judgement*, tr. J. C. Meredith (Oxford, 1928), 31.
[3]*Ibid.*

etc. Only a clear conception of the nature of the different sciences can eliminate confusions of this kind. "It is sometimes difficult," says Kant, "to define what is meant by a science. But science gains in precision by the establishment of a definite conception of it, and many errors from different sources are thus avoided which otherwise slip in when we are unable to distinguish the science from the cognate sciences."[4]

It is, therefore, for Kant one of the functions of philosophy to determine the scope of the various sciences. He considered it to be "of the utmost importance to *isolate* the various modes of knowledge according as they differ in kind and origin, and to secure that they are not confounded owing to the fact that usually, in our employment of them, they are combined. What the chemist does in the analysis of substances, and the mathematician does in his special disciplines, is in still greater degree incumbent upon the philosopher, that he may be able to determine with certainty the part that belongs to each special kind of knowledge in the diversified employment of the understanding and its special value and influence."[5]

The sciences do not exist with their definitions ready to hand. Although a particular science may be fully established it is still not an easy task to define it, and certainly no science begins historically with a clear or adequate definition of itself. The historic originator of any science must, of course, have some vague idea of it, otherwise he would not be able to set it on its way. This vague idea of a science in its early stages is described by Kant as something which lies hidden in reason "like a germ in which the parts are still undeveloped and barely recognizable even under microscopic observation."[6] Hence when looking for the definition of a science we must not adopt the description which is given by the founder of that science. The natural unity of the parts of the science only emerges as the science develops. It is to the principle of this natural unity that we must look for the idea of the science, not to what the father of the science says. "For we shall then find that its founder, and often even his latest successors, are groping for an idea which they have never succeeded in making clear to themselves, and that consequently they have not been in a position to determine the proper content, the articulation (systematic unity), and limits of the science."[7] In the initial stages of

---

[4]*Introduction to Logic,* tr. T. K. Abbott (London, 1885), 12.
[5]*Critique of Pure Reason,* tr. N. K. Smith (London, 1933), A 842 = B 870.
[6]*Ibid.,* A 834 = B 862.   [7]*Ibid.*

the development of a science materials are collected together in random fashion, but none the less under the direction of a vague idea. This directing idea attains clarity gradually as over a long period more and more materials are amassed and begin to assume the character of a system. "Systems seem to be formed in the manner of lowly organisms, through a *generatio aequivoca* from the mere confluence of assembled concepts, at first imperfect, and only gradually attaining to completeness, although they one and all have had their schema, as the original germ, in the sheer self-development of reason."[8]

In the purposes determining the development of a science Kant distinguishes those which are occasioned by the purely contingent circumstances of the individual inquirer, i.e., his private reasons, and those which are based on a certain "universal interest." Reason has essential ends of its own which express themselves in the same way for all inquirers, and it is these ends which give to the sciences their systematic character, or what Kant calls "architectonic unity." The kind of unity based on the individual's historically conditioned purposes is only a "technical unity," and it is not to this that we must look in order to attain the conception of a science.

The one science for which Kant was above all concerned to reach a definition, or clear conception, was metaphysics, and his discussion of this matter serves to illustrate the points which have just been made. He makes two apparently contradictory statements: (1) "The idea of such a science is as old as speculative human reason," and (2) ". . . the philosophers failed in the task of developing even the idea of their science. . . ."[9] On the one hand the disposition to Metaphysical speculation is intrinsic to human reason. Man as rational is compelled to take an interest in his ultimate destiny, and there is no rational being who does not speculate about what lies beyond sensible experience. "Indeed," says Kant, "we prefer to run every risk of error rather than desist from such urgent inquiries on the ground of their dubious character, or from disdain and indifference. These unavoidable problems set by pure reason itself are *God, freedom,* and *immortality.* The science which with all its preparations, is in its final intention directed solely to their solution is metaphysics. . . ."[10] But though the idea of such a science arises directly out of a universal interest of

[8]*Ibid.*, A 835 = B 863.
[9]*Ibid.*, A 842 = B 870; A 844 = B 872.
[10]*Ibid.*, B 7.

reason, and inevitably directs inquiry towards a set of problems which Kant calls metaphysical, because they transcend the realm of sensory experience, nevertheless the idea of the science is still only a vague one. An exact distinction has never been made between what belongs within metaphysics and what does not.

Kant cites two confusions which have served to prevent the attainment of a clear view of metaphysics, and his account implies in addition a radical criticism of the conceptions entertained by Bacon, Descartes, and Leibniz of the unity of the sciences. The first of these confusions is found in the definition of metaphysics as the science of the first principles of human knowledge. Such a definition fails to distinguish metaphysics as a special *kind* of knowledge and gives it only "a certain precedence in respect of generality." If knowledge is ordered only with respect to degrees of generality of principles, then one may ask where the dividing line is to be drawn between principles which are to be considered first and those which are to be considered subordinate. "Does the concept of extended body belong to metaphysics? You answer, Yes. Then, that of body too? Yes. And that of fluid body? You now become perplexed; for at this rate everything will belong to metaphysics. It is evident, therefore, that the mere degree of subordination (of the particular under the general) cannot determine the limits of a science; in the case under consideration, only complete difference of kind and of origin will suffice."[11]

This mode of determining the nature of metaphysics in terms of its degree of generality was used by Bacon, but Kant's criticism applies not only to Bacon's metaphysics, but also to his conception of the unity of natural history, physics, and metaphysics in a pyramid of knowledge, in which there is an unbroken and uniformly graded ascent from particulars to the summary law of nature. Whatever unity Kant conceives the sciences to have it will not be of that kind, for in it there can be no real distinctions among the sciences. It would remain a matter of arbitrary decision to say where one science begins and another ends.

The other source of confusion arises from the seductive influence exercised by mathematics upon philosophers. If metaphysics and mathematics have this in common, that both are *a priori* in their origin, then it is very natural to suppose that metaphysics can attain

[11]*Ibid.*, A 844 = B 872.

the same remarkable successes that have characterized the advance of mathematics, by adopting the method of the latter. The assimilation of philosophy to the type of mathematics is the subject of Kant's criticism in the longest and perhaps the most important section of his Transcendental Doctrine of Method. It is significant with reference not only to Descartes' and Leibniz's conceptions of metaphysics, but also to their conception of the unity of the sciences, in accordance with which the different sciences can be ordered in relation to one another in the same way as the truths of mathematics. If philosophy is a different kind of knowledge from mathematics, then the principle of organization relating the sciences to one another cannot be the same as that found governing the parts of a mathematical system, for the ordering of the sciences is for Kant a philosophical task and must consequently proceed in accordance with the method of philosophy. The sciences taken together as an organized whole will not form a single axiomatic or deductive system.

Kant's insistence on a clear-cut division of the sciences is complemented by his equal insistence that all the sciences taken together have the unity of a single organized whole. Kant's definition of science is that it is any doctrine constituting a system: " . . . systematic unity is what first raises ordinary knowledge to the rank of a science, that is, makes a system out of a mere aggregate of knowledge. . . ."[12] In common knowledge, as opposed to scientific knowledge, there is a mere aggregate of cognitions and the parts precede the whole; whereas in scientific knowledge the system comprising the science rests on the idea of the whole which precedes the parts. Historically, Kant's definition marks a new departure. It is also, as Lalande points out, the one that is classic today.[13] The Cartesian definition which identifies science with apodictic certainty, and the Aristotelian definitions of science as knowledge of causes, and knowledge of the necessary, are not abandoned by Kant, but they are made subordinate to the requirement of systematic unity as the principal distinguishing mark of science. The Cartesian definition of science as "true and evident cognition" is, where Kant is prepared to be strict, retained within his own definition. Thus he says, "That can only be called science *proper* whose certainty is apodictic: cognition that can merely contain

[12]*Ibid.*, A 832 = B 860.

[13]A. Lalande, "Science," *Vocabulaire technique et critique de la philosophie* (Paris, 1951).

empirical certainty is only improperly called science."[14] But the possession of systematic character is so much more important than certainty of cognition, that Kant allows that we may speak of natural *science*, even though natural science contains laws which are empirical, and which do not therefore possess apodictic certainty. Nevertheless, where the knowledge of nature, including what is empirical, does not rest ultimately on purely rational, and therefore apodictically certain, principles, Kant will not extend the name "science" to it. For science, as a connection of cognitions in a system, is, he says, "a system of causes and effects." That is to say, it is, in accordance with a classic definition of science, knowledge of what is *necessary*. If the principles of a systematic body of cognitions

. . . are in the last resort merely empirical, as, for instance, in chemistry, and the laws from which the reason explains the given facts are merely empirical laws, they then carry no consciousness of their *necessity* with them (they are not apodictically certain), and thus the whole does not in strictness deserve the name of science; chemistry indeed should be rather termed systematic art than science. . . . As the word nature itself carries with it the conception of law, and this again the conception of the *necessity* of all the determinations of a thing appertaining to its existence, it is easily seen why natural science must deduce the legitimacy of its designations only from a *pure* part of it, namely, that which contains the principles *a priori* of all remaining natural explanations; and why only by virtue of this portion it is properly science, in such wise, that, according to the demands of the reason, all natural knowledge must at last turn on natural science and there find its conclusion.[15]

[14]*Metaphysical Foundations of Natural Science*, tr. E. B. Bax (London, 1891), 138.

[15]*Ibid.*, 138 f. For Kant there can be a natural science proper only to the extent that mathematics is applicable to the kind of natural phenomena with which the science specifically deals. By mathematics it is possible to construct the spatial intuition corresponding *a priori* to the empirical concept of corporeal nature which we get through our outer senses. Consequently mathematics can provide the *a priori* basis necessary for a science of matter. With the natural phenomena of inner sense, however, mathematical principles are not applicable, for the pure intuition which we construct to correspond to them is in time only and not in space. Empirical psychology "can never therefore be anything more than an historical, and as such, as far as possible systematic natural doctrine of the internal sense, i.e. a natural description of the soul, but not a science of the soul. . . ." In chemistry it is impossible by mathematics to construct a conception for the chemical interactions of substances which is capable of being presented *a priori* in intuition. Hence the principles of chemistry, too, remain merely empirical. Chemistry has, however, the advantage over psychology of being experimental, for substances can be joined and separated at will, whereas psychological phenomena cannot. Moreover, internal observation alters the character of what is observed. There can therefore be no "psychological experimental doctrine." *Ibid.* 140–2.

There are, then, two elements in Kant's conception of the nature of science: (1) it is a systematic unity of cognitions, and the name science may be extended even to such systems as include empirical, and therefore not apodictically certain, knowledge; (2) such a system must, however, even if it includes empirical knowledge, rest on principles which are themselves necessary and certain. An organized body of merely empirical knowledge would not be science, nor at the same time would knowledge or causes, or what is necessary, be in itself sufficient to constitute a science, for common knowledge can be knowledge of what is necessary, yet as unsystematic it remains unscientific. Two things appear to follow from the definition of science as any doctrine constituting a system. First, the total system of the sciences would itself be a single science, containing the other sciences as the subordinate parts of an organic whole. These subordinate sciences might in turn be comprised of still lesser wholes. Secondly, the logical principles determining the unity of the sciences in a single system would not be different in character from those determining the unity of cognitions in any one of the sciences taken individually.

THE UNITY OF NATURE AND THE UNITY OF SCIENCE

The unity of science has its foundations for Kant in the nature of reason itself. It is not, as with Bacon, the reflection of a unity found in nature. Its origin is purely subjective; it is prescribed *a priori* to the objects of science, not empirically determined. In speaking of the unity of nature and the relation of the unity of science to it, it is necessary, however, to keep distinct two ways in which Kant, in different contexts, refers to the unity of nature. In the first place Kant refers to nature as a unity in so far as it is a system of necessarily interconnected phenomena. Unless the sum total of phenomena comprised such a system there would be no nature at all, either for ordinary knowledge or for scientific knowledge. The laws governing these necessary connections and giving nature its unity are imposed by the understanding upon the manifold of appearances, and they have their source in the synthetic unity of apperception. The necessary connections are those of substance and accident, cause and effect, and the interaction of substances. The three principles, the "analogies of experience," which express these connections, taken together "declare," says Kant, "that all appearances lie, and must lie, in *one* nature, because without this *a priori* unity, no unity of experience, and therefore no determination of objects in it, would be possible."[16]

Within this one nature of necessarily interconnected appearances experience reveals an endless variety of substances, causal sequences, and causal interactions between substances. What distinguishes scientific knowledge from common knowledge is the effort made by the scientist to reduce this multiplicity to unity so far as he is able, by bringing particular substances under species, species under genera, and these in turn under still higher genera, thereby showing in what way the differences among them are merely variations of the same fundamental substance. Similarly with the infinite variety of ways in which phenomena are causally determined, the scientist will seek to reduce the multiplicity of causal laws to an ever smaller number of laws. In proceeding in this fashion the scientist is assuming that there is in nature an identity underlying the seeming heterogeneity. This second conception of the unity of nature in terms of the identity underlying differences is the one which is relevant to the discussion of the unity of science as contrasted with ordinary knowledge. Without the first kind of unity there could, of course, be no nature at all for the scientist to investigate. It is constitutive of nature. But the second kind of unity is not regarded by Kant as constitutive of nature, but merely as regulative of scientific inquiry, and he denies to it the objective validity which must be accorded to the first. It is true that the scientist is compelled to impute it to nature, whether he realizes that he is doing so or not, for if science is to have the unity which reason by its very nature demands, he must suppose that this unity actually exists in nature, and that it is there to be discovered. But he cannot empirically derive his conception of it from nature, for since this conception directs all his inquiries into nature, and indeed gives rise in the first place to those inquiries, it must precede them. If science, then, in contrast with ordinary knowledge, is a systematic unity of cognitions, the nature of its unity is not derived empirically from nature, but is antecedently determined by a certain demand of reason. It is therefore to the nature of reason that we must look for the basis of the unity of science.

[16]*Critique of Pure Reason*, A 216 = B 263.

In his theory of the nature of reason Kant assigns to it two distinct functions. One is the merely formal or logical use of reason, by which it draws mediate inferences. In its logical use reason abstracts from all content of knowledge, and in this respect stands contrasted with its other use, which is, or at least claims to be, cognitive, and which Kant calls its "real" or "transcendental" use. Here reason is the source of certain concepts or principles which it possesses independently of experience. Of these two it is the logical employment of reason which is initially decisive for Kant's theory of the unity of science. He asserts that there are three essential elements in all inference:

(1) a universal rule, which is entitled the major premise;

(2) the proposition which subsumes a cognition under the condition of the universal rule, and which is entitled the minor premise; and lastly,

(3) the conclusion, the proposition which asserts or denies of the subsumed cognition the predicate of the rule.

Thus, according to this theory, the universal rule, or major premise, connects the predicate with a condition. The minor premise states that its subject fulfills the condition, and in the conclusion the subject is determined by the predicate which has been connected with the condition.[17] Where the conclusion is something still to be established there I have to discover whether or not it stands under certain conditions according to a universal rule.

This account of what takes place in inference shows, Kant says, what reason in its logical employment is aiming at. It is trying to reduce the variety and multiplicity of knowledge got through the understanding to the smallest number of principles or universal conditions, and in that way to give it its highest possible unity. This unity is its necessary goal. It must by its very nature try to advance towards it.

. . . reason, in its logical employment, seeks to discover the universal con-

---

[17]To contrast Kant's principle of the syllogism with the *dictum de omni* or the *nota notae* it can be given Joseph's conveniently succinct formulation: "Whatever satisfies the condition of a rule falls under the rule." He illustrates the principle in the following way: "In the rule 'Whatever is *B*, is *A*', being *B* is the condition, the fulfilment of which involves being *A*; and to a given concept *C* fulfilling the condition the rule will apply, and it will be *A*." H. W. B. Joseph, *An Introduction to Logic* (Oxford, 1916), 309. Kant intended his principle to apply to the disjunctive and hypothetical syllogisms as well as to the categorical.

dition of its judgment (the conclusion), and the syllogism is itself nothing but a judgment made by means of the subsumption of its condition under a universal rule (the major premise). Now since this rule is itself subject to the same requirement of reason, and the condition of the condition must therefore be sought (by means of a prosyllogism) whenever practicable, obviously the principle peculiar to reason in general, in its logical employment, is:—to find for the conditioned knowledge obtained through the understanding the unconditioned whereby its unity is brought to completion.[18]

In acting in accordance with this principle of its logical employment, reason is entirely non-cognitive. It is the understanding in conjunction with sensibility which alone gives rise to knowledge. It is reason which demands that this knowledge be brought into systematic unity. Its sole function is "to prescribe to the understanding its direction toward a certain unity of which it itself has no concept, and in such a manner as to unite all the acts of the understanding in respect of every object, into an *absolute* whole."[19] It is important to emphasize that the unity of science is a purely methodological and subjective principle arising solely out of reason in its logical employment. We have no basis whatever for asserting that such a unity exists in the objects of our knowledge. It is necessary to emphasize this because reason compels us to think of this unity as existing objectively. We are not free to entertain the supposition that possibly nature is ultimately heterogeneous, and may not possess a systematic unity. "Reason would then run counter to its own vocation. . . . The law of reason which requires us to seek for this unity is a necessary law, since without it we should have no reason at all, and without reason, no coherent employment of the understanding, and in the absence of this, no sufficient criterion of empirical truth. In order, therefore, to secure an empirical criterion we have no option save to pre-suppose the systematic unity of nature as objectively valid and necessary."[20]

The systematic unity of nature has not always been acknowledged by scientists, but nevertheless, says Kant, we find it "covertly implied, in remarkable fashion, in principles on which they proceed." Here Kant enunciates three principles which govern scientific inquiry and which rest on this supposition of the unity of nature.[21] First, there is

[18]*Critique of Pure Reason*, B 364.   [19]*Ibid.*, B 383.
[20]*Ibid.*, A 651 = B 679.
[21]In the *Critique of Judgement* these principles are attributed to the reflective judgment.

the principle that principles must not be unnecessarily multiplied—
entia praeter necessitatem non esse multiplicanda. This principle
presupposes that unity is to be found in nature; that behind the infinite variety of nature there is "a unity of fundamental properties—
properties from which the diversity can be derived through repeated
determination." There can be no reasoning at all unless it is supposed
that among individuals there is the identity of species, and that the
species are only determinations of certain genera, and these in turn
of still higher genera, and that, in short, our concepts have a systematic unity; for reasoning consists in concluding from the universal
to the particular, and that is only possible in so far as universal properties are ascribed to things as being the foundations upon which the
particular properties rest.

This demand for identity which governs scientific inquiry can be
illustrated in chemistry.[22] Chemistry made a great advance, says
Kant, when all salts were reduced to two main genera, acids and
alkalis. Chemists found themselves, however, still compelled to try
to show that the difference between these two is only a variation of
one basic material. The same attempt has gone on to find a basic
identity in the different kinds of earths, but, even more, the chemists
are unable to banish the thought that there is even a common principle for both the earths and the salts.

Kant insists that this principle—"the logical principle of genera,
which postulates identity"—is not regarded by the scientist merely as
a device for economy in explanation to save himself unnecessary
trouble. He must think of it as having objective validity. The unity in
variety is not supposed as an hypothesis, which may prove to be
successful. Reason does not ask us to try it out, but insists that the
unity is there to be found; it "does not here beg but command."[23]

The logical principle of genera is balanced by a second principle,
that of species, to which Kant gives the formulation, *entium varietates
non temere esse minuendas.* Where the first principle compels the
scientific inquirer to seek identity, the second compels him to seek
diversity. The aim of the first in ascending to the genus is to secure
unity in the system of our knowledge. The aim of the second in des-
cending from the genus is to secure completeness in the system. It
sets a goal which, of course, can never be reached, for in the descent

[22]However, see n. 15 on the dubious status of chemistry as a science.
[23]*Critique of Pure Reason,* A 653 = B 681.

from genera to species no species can be regarded as the lowest. "For since the species is always a concept, containing only what is common to different things, it is not completely determined. It cannot, therefore, be directly related to an individual, and other concepts, that is, subspecies, must always be contained under it."[24] Although the principle of specification is merely a logical principle governing scientific inquiry, nevertheless, as with the principle of parsimony, the inquirer is compelled to regard it as having its foundation in nature. To return again to chemistry, the discovery that absorbent earths are of different kinds could only have been possible if we made the prior assumption that such differences do actually exist in nature and are there to be discovered. This assumption functions as a rule directing the understanding to the task of seeking out these differences, and without such an assumption there would be nothing to occasion the exercise of the understanding in scientific inquiry.

The third logical principle directing inquiry towards the goal of systematic unity prescribes that we proceed from each species to every other by gradual increases in degrees of diversity. This principle of the continuity of forms arises out of the union of the other two, for the systematic unity reached by the ascent to higher genera and the descent to lower species requires that the progression be unbroken. It compels us to deny that there can be a plurality of original first genera existing in isolation from one another, "separated, as it were, by an intervening empty space." We have to suppose that the various genera are simply divisions of one single highest and universal genus. This implies that there are no leaps in the transition from one species to another, but that the differences between species are mediated by intervening smaller differences. As with the other two, this principle, although it has its origin solely in the logical employment of reason, cannot be regarded by us as merely a methodological device or as a useful hypothesis. We have to think of this affinity of forms as actually existing in nature.

THE "IDEA" OF A SCIENCE

These three logical principles, by which all scientific inquiry is regulated, show that the attitude of reason to any body of knowledge is to secure its systematization, i.e., to show the interrelation of its

[24]Ibid., A 656 = B 684.

parts in accordance with a single principle. To achieve that end for any particular science, reason must presuppose an idea of the form of the science as a whole, a whole which is prior to the parts, and which contains the conditions that determine *a priori* the position of each part and its relation to the other parts. It is only if such an idea is presupposed that a body of knowledge can constitute a science, that is to say, be a system and not a mere aggregate of knowledge. This presupposed idea, however, like the three principles enunciated above, has no objective validity. Although reason inevitably hypostasizes it, its sole significance is that it represents the complete unity of the science for the purpose of directing the understanding. No empirical derivation is possible for the idea of the science, for the empirical investigation of nature gets its very direction from the idea. Although the idea appears to signify something objective, it is in reality only a schema—or more accurately the analogon of a schema— by which reason can represent the systematic unity it demands for our empirical knowledge.

In calling an idea the analogon of a schema Kant is attributing to the idea a function in the organization of a science analogous to that performed by the schematism of the understanding in bringing the manifold of intuitions under the unity of concepts. The concepts of the understanding and sensible intuitions are heterogeneous. If sensible appearances are to be brought under the categories, some factor is required to mediate between the two, something which is homogeneous on the one hand with the category and on the other with the appearances. It is the transcendental imagination which performs this task of mediation by providing a schema of the concept, and thereby making possible the application of the concept to what is given in intuition.

The unity of a science stands in the same relation to the knowledge provided by the understanding as the concepts of the understanding do to the manifold of appearances. Like a category, the unity of a science is, when taken in itself, undetermined—it is simply the unity of a manifold *in general*, and as such empty and meaningless. Just as the schemata provide the rules of procedure for unifying the manifold of intuitions under the categories, so the ideas provide the rules for bringing the manifold of knowledge provided by the understanding into systematic unity, thereby transforming this knowledge into *scientific* knowledge. The idea may therefore be called the analogon of a schema. It is, however, only an analogon, and not an actual schema. No sensible schemata can be supplied by the imagination for the unity of science, for this unity is incapable of having any object in sense experience corresponding to it.

But if it is not possible to provide a schema for the unity of reason, it is, nevertheless, possible to symbolize this unity, or represent it indirectly by means of an analogy with what is given in experience. By symbolizing the unity we acquire a rule of procedure for attaining it. This may be illustrated in connection with the last and highest degree of formal unity under which natural science is brought, namely, the purposive unity of things. This unity is symbolized by means of an analogy with human art, the latter being something with which we have acquaintance in experience. By means of this analogy nature as a whole can be regarded as the work of a supreme intelligence. This idea of God will now serve to direct scientific inquiry, by providing a rule of procedure for attaining the highest degree of systematic unity. It is able to function as a rule, because we shall then proceed in our inquiries *as if* the world were the work of a supreme artist. Unless the unity of a science is in some way symbolized by an idea, scientific inquiry remains undirected. It is not enough for reason simply to demand unity in what is known, for "the *unity of reason* is in itself *undetermined*, as regards the conditions under which, and the extent to which, the understanding ought to combine its concepts in systematic fashion."[25]

The question of the rôle of reason in the scientific organization of knowledge is raised by Kant in connection with his chief problem, the possibility of a science of metaphysics. The principal branches of metaphysics in the scheme he took over from Wolff were psychology, cosmology, and rational theology. Kant's immediate concern with the function of reason was to show how, in fulfilment of its purely logical employment, reason gives rise inevitably to three transcendental ideas of the unconditioned, corresponding to each of the three kinds of syllogism, categorical, hypothetical, and disjunctive. The three transcendental ideas are (1) the idea of the absolute (unconditioned) unity of the thinking subject, (2) the idea of the absolute unity of the series of conditions of appearance, and (3) the idea of the absolute unity of the condition of all objects of thought in general.

"The thinking subject is the object of *psychology*, the sum total of

[25]*Ibid.*, A 665 = B 693.

all appearances (the world) is the object of *cosmology*, and the thing which contains the highest condition of the possibility of all that can be thought (the being of all beings) is the object of *theology*. Pure reason thus furnishes the idea for a transcendental doctrine of the soul (*psychologia rationalis*), for a transcendental science of the world (*cosmologia rationalis*), and, finally, for a transcendental knowledge of God (*theologia transcendentalis*)."[26] The whole argument of the Transcendental Dialectic is designed to show that these sciences are pseudo-sciences, and that the true function of the idea of the soul, of the world as a whole, and of an *ens realissimum*, is to direct the understanding in scientific inquiry into nature. They are ways in which reason pursuing merely its logical function prescribes a systematic unity for our knowledge. The implication of this doctrine is that the only theoretical science is natural science, and that this has two branches, psychology and physics, the one regulated by the idea of the unconditioned thinking subject, the other by the idea of the cosmos, and that these two sciences are finally unified in a single science under the transcendental idea of God.[27]

A difficulty, however, emerges here in connection with Kant's use of the word "idea." Although the three ideas of the Dialectic are schemata for the three *natural* sciences of psychology, physics, and the science which is a systematic union of psychology and physics, nevertheless they are not conceptions of these sciences, and we do not look to them for the definitions of these sciences. On the contrary they are conceptions of three *metaphysical* sciences, rational psychology, rational cosmology, and transcendental theology, which are, moreover, pseudo-sciences. Thus it would appear that "idea" in the sense in which it is used by Kant to signify the schematic representation of the systematic unity of a science, and "idea" as synonym for "conception" of a science are not equivalent uses of the word in the Dialectic. The idea of cosmology is the conception of cosmology, but it is the schema for physics. In the Transcendental Doctrine of Method, on the other hand, the idea of a science signifies both the conception of it (or the basis for its definition) and the schema for its systematic unity, as is indicated clearly in Kant's statement that until

[26]*Ibid.*, B 391 = A 334.

[27]In the *Metaphysical Foundations of Natural Science* Kant denies that psychology can ever be scientific. See n. 15. Hence the term "natural science," used strictly, is, he says, applicable only to the science of corporeal nature.

the idea is made clear the scientist is not "in a position to determine the proper content, the articulation (systematic unity), and limits of the science."[28]

THE SYSTEMATIC UNITY OF ALL KNOWLEDGE IN RELATION TO THE ULTIMATE END OF HUMAN REASON

Kant has accounted for the origin and development of the natural sciences, as well as their unity in one comprehensive science of nature, in terms of an interest of reason, an interest which belongs to reason in its logical employment. But reason also has certain moral interests which belong to it in its practical employment as governor of the will. The pursuit of philosophy with a view to the attainment of systematic unity in its merely logical perfection, Kant assigns to the "scholastic concept" of philosophy.

"But there is likewise another concept of philosophy, a *conceptus cosmicus*, which has always formed the real basis of the term 'philosophy', especially when it has been as it were personified and its archetype represented in the ideal *philosopher*. On this view, philosophy is the science of the relation of all knowledge to the essential ends of human reason (*teleologia rationis humanae*), and the philosopher is not an artificer in the field of reason, but himself the lawgiver of human reason."[29] The mathematician, the physicist, and the logician are declared to be only artificers in the field of reason, since they merely seek the systematic unity within their respective sciences which is required by reason in its logical employment, although reason employs them all the time in the fulfilment of its essential ends without their knowledge. It is necessary, then, to determine what philosophy, according to the cosmic concept of it, prescribes concerning the systematic unity of all knowledge in relation to these ends.

It is characteristic of reason, not only in its logical employment, but

[28]*Critique of Pure Reason*, A 834 = B 862. To define a science is, for Kant, "to determine accurately that peculiar feature which no other science has in common with it, and which constitutes its specific characteristic. . . . The characteristic of a science may consist of a simple difference of *object*, or of the *source of knowledge*, or of the *kind* of knowledge, or even perhaps of all three together. On this characteristic, therefore, depends the idea of a possible science and of its territory." *Prolegomena zu einer jeden künftigen Metaphysik, die als Wissenschaft wird auftreten können* (Riga, 1783), § 1.
[29]*Critique of Pure Reason*, A 839 = B 867.

also in reflecting on its own essential ends, to demand complete systematic unity. In the case of the logical employment of reason Kant was at pains to show how this demand arises from a certain interest of reason. But if one asks why reason demands a systematic unity of its own ends or interests, Kant is no longer concerned to explain this demand in terms of an interest. The demand that principles should not contradict one another is "necessary for the possibility of any employment of reason at all," and "constitutes no part of its interest."[30] Reason would not be reason if in reflection on its own essential ends it did not seek their unity. If the essential ends of reason have a systematic unity, then one of them, Kant argues, must be an ultimate end and the others be subordinate to it as means. This ultimate end of reason is "no other than the whole vocation of man, and the philosophy which deals with it is entitled moral philosophy."[31] If the highest aims of reason have a systematic unity, then we may suppose that the sciences which exist in order to fulfil these aims must also have a systematic unity, and that therefore both theoretical knowledge and practical knowledge must ultimately belong together in one philosophical system. It is to this final synthesis of knowledge that we now turn.

Reason in its speculative employment was shown in the *Critique of Pure Reason* to be by its very nature compelled to seek the unconditioned for the conditioned knowledge which it obtains through the understanding. In its practical employment reason is governed by the same necessity. "As pure practical reason, it likewise seeks to find the unconditioned for the practically conditioned (which rests on inclinations and natural wants), and this not as the determining principle of the will, but even when this is given (in the moral law) it seeks the unconditioned totality of the *object* of pure practical reason under the name of the *Summum Bonum*."[32]

In the Analytic of the Pure Practical Reason and in the *Grundlegung* Kant had argued that the good will is the only thing which is unqualifiedly good, and that all the things which appear most desirable and which go to make up happiness, can only be considered good conditionally upon their being found united with a good will. The good will, i.e., a will determined by the moral law, is the indispensable

[30]*Critique of Practical Reason*, tr. T. K. Abbott (London, 1909), 216.
[31]*Critique of Pure Reason*, A 840 = B 868.
[32]*Critique of Practical Reason*, 203.

condition of being worthy of happiness. But while the good will or virtue is the *supreme* good, it is not by itself the whole or perfect good. Persons are ends in themselves, and if they were to deserve happiness, but yet not enjoy it, the situation would be one which falls short of the whole or perfect good. Hence Kant concludes that it is virtue and happiness taken together which constitute the *summum bonum* of the individual person, and "the distribution of happiness in exact proportion to morality (which is the worth of the person, and his worthiness to be happy) constitutes the *summum bonum* of a possible world."[33]

This peculiar characteristic of the *summum bonum*, that it is made up of two distinct but necessarily connected elements, virtue and happiness, has profound consequences for Kant's ultimate synthesis of theoretical and practical knowledge in a unified system. Man finds himself in a situation in which only one of the two elements of the *summum bonum*, virtue, lies within his power, for the goodness of the will is not affected by the success or failure of its actions. The other element in the *summum bonum*, namely happiness or the state of affairs in which everything goes according to man's desires, does not lie within his powers, but is dependent upon the course of nature. Man, though acting rationally in the world, "is not the cause of the world and of nature itself." We have, nevertheless, a duty to promote the *summum bonum*. Only that can be a duty which is possible, and therefore the attainment of the *summum bonum* must be possible. It is possible, however, only on the condition that God exists. It becomes, then, "morally necessary to assume the existence of God." There must be a being distinct from nature itself, capable of harmonizing the whole physical order with man's ends.

Here it is necessary to refer once more to the function of ideas in furnishing the conception of a science and in regulating the pursuit of that science. The idea is a schema, or rather the analogon of a schema, for the systematic ordering of what belongs within the science. Both natural science and morality are governed by an idea. In the case of natural science the last and highest degree of formal unity is "the purposive unity of things." The demand of reason for unity compels it "to regard all order in the world as if it had originated in the purpose of a supreme reason. Such a principle opens out to

[33]*Ibid.*, 206.

our reason, as applied in the field of experience, altogether new views as to how things of the world may be connected according to teleological laws, and so enables it to arrive at their greatest systematic unity."[34] This idea of a teleological system of nature governed by a supreme intelligence has its counterpart in morality in the idea of a Moral World, or Kingdom of Ends. The Kingdom of Ends is the union in a systematic whole of all rational beings, considered as ends in themselves, together with the special ends which each individual proposes to himself. This conception of a Kingdom of Ends functions as a practical idea. It provides a means of formulating the moral law in such a way as to bring it "nearer to intuition (by means of a certain analogy), and thereby nearer to feeling."[35] As the analogon of a schema it supplies an "as if" mode of formulating the moral law for directing action in accordance with that law. The law can be stated thus: "Every rational being must so act as if he were by his maxims in every case a legislating member in the universal kingdom of ends."[36]

The problem originally stated in terms of the *summum bonum* can now be restated in terms of the kingdom of ends. The Kingdom of Ends would be capable of realization only if, in addition to men's acting in accordance with the moral law, nature also harmonized with human ends, i.e., were teleologically organized with respect to those ends. This in turn could only be possible if both the kingdom of nature and the kingdom of ends belonged together in a single system under one supreme ruler. Thus the inescapable fact of duty involves the necessity of postulating the existence of God.

In this moral theology there resides the unity of theoretical and practical knowledge. Reason in its theoretical employment compels us to regard nature as a single teleologically organized system, but it is only moral theology which can show what final end this system is teleologically directed towards, namely, man as a moral being. This unity of theoretical and practical knowledge is not, however, one which is of the slightest significance for natural science. It is significant only for morality. The practical reason, however successful it may be in establishing that which lies beyond the competence of theoreti-

---

[34]*Critique of Pure Reason*, A 687 = B 715.
[35]*Fundamental Principles of the Metaphysic of Ethics*, tr. T. K. Abbott (London, 1946), 65.
[36]*Ibid.*, 68.

cal reason, nevertheless does not lend its principles to the directing of natural science. Kant is emphatic on this point. The conception of man in his moral capacity as the end of physical nature cannot be applied to the cognition of nature. "*The only possible use of this conception* is for practical reason according to moral laws."[37] Speculative reason is able to pursue its ends without borrowing any concepts from the practical reason. Practical reason, on the other hand, is only able to fulfil its end by postulating the unity of the world of nature and the moral world under a supreme ruler. The conception of nature, and in particular a nature exhibiting purposive unity is contributed to the practical reason by the theoretical judgment. Without this contribution, moral philosophy would remain incomplete.

To say, however, that natural science contributes to the fulfilment of the ends of practical reason, is to say that it derives its ultimate significance for philosophy from the fact that it does make this contribution. The natural scientist, seeking only the logical perfection of his science, is, Kant has said, a mere artificer. The *conceptus cosmicus* of philosophy requires us to view all knowledge in terms of its relation to the essential ends of human reason. From this point of view the natural scientist is seen to be merely an instrument, however unwitting, in the furthering of these ends.

Kant's doctrine may now be summed up. All systematic unity takes place in accordance with ends of reason. It is always an idea which directs this unification. As we have seen, the various ends of reason must themselves possess a systematic unity under one ultimate end —"the whole vocation of man." Such a unity, like all others, is directed by an idea—the moral idea of God, and the sciences which answer to the various ends of reason attain a corresponding unity under moral theology, for it is moral theology which, Kant says, "enables us to fulfil our vocation"[38] or attain our highest end. Thus the ultimate unity of the sciences is attained for Kant in moral theology.

[37]*Critique of Teleological Judgement,* 124 (italics mine).
[38]*Critique of Pure Reason,* A 819 = B 487.

Appendix

Descartes on *Metaphysics as Science of the Principles of Knowledge*

As the science of the principles of knowledge, metaphysics gives "the explanation of . . . all the clear and simple notions which are in us."[1] These simple or "primitive notions," Descartes tells Princess Elizabeth, are "very few in number; for after the most general, being, number, duration, etc., which are common to everything we can conceive, we have for body in particular, only the notion of extension, from which follow those of figure and motion; and for the soul taken by itself, we have only that of thought, in which are included the perceptions of the understanding, and the inclinations of the will; finally for the soul and body taken together, we have only the notion of their union, on which depends that of the force with which the soul moves the body, and the body acts on the mind, causing its feelings and passions."[2] Of all these primitive notions, it is the last three which are by far the most important in Descartes' metaphysics; namely, thought, extension, and the substantial union of the two in man's nature. The last is a genuinely primitive notion, and it cannot be arrived at by compounding the other two primitive notions. All the simple notions listed above are of things and their qualities. In addition to them there are certain "common notions or axioms," such as the principle, *ex nihilo nihil fit*, or the principle, it is impossible that the same thing can be and not be at the same time. There are many such axioms, and Descartes confesses to having no interest

[1] *Principles of Philosophy*, Preface, *The Philosophical Works of Descartes*, tr. E. S. Haldane and G. R. T. Ross (Cambridge, I, 1931; II, 1934), I, 211.
[2] *Letter*, 21 May, 1643, *Descartes, Correspondance*, ed. C. Adam and G. Milhaud (Paris, 1936 *et seq.*), V, 290.

in making an exhaustive list of them. He expresses contempt for the kind of metaphysics like Aristotle's which makes the principle of contradiction the first principle of metaphysics; Descartes seeks an entirely different kind of first principle—one asserting existence.

The word *principle* can be taken in different senses. . . . It is one thing to seek *a common notion*, which is so clear and so general that it can be used as a principle for proving the existence of all the beings, the *Entia*, which can later come to be known; it is another thing to seek *a being*, the existence of which is more known than that of any other beings, so that it can be used as a *principle* for coming to know them.

As used in the first sense, it can be said that *impossibile est idem simul esse et non esse* is a principle, and that it can generally serve, not properly for causing us to know the existence of anything, but only, once we do know it, for providing confirmation of its truth by the following kind of reasoning: *it is impossible that that which is should not be; now I know that such and such a thing is; therefore I know that it is impossible that it should not be.* This is of very little importance, and it makes us none the wiser.

In the other sense, the first principle is that *our soul exists*, since there is nothing whose existence is better known.

Descartes adds that it is not a condition of this latter kind of first principle that all other propositions can be proved by means of it. It is sufficient that it can be used for finding several other truths, and that there are no others on which it depends, or which can be more easily discovered. "For it may well be that there is nowhere in the world any one principle to which alone everything can be reduced; and the way in which other propositions are reduced to the principle: *impossibile est idem simul esse et non esse,* is superfluous and useless; whereas it is with very great usefulness that we begin by assuring ourselves of the *existence of God,* and then the existence of all his creatures, *by the consideration of our own existence.*"[3]

"I am, I exist" is the first principle of Descartes' metaphysics, as presented in the *Meditations,* a work which employs his method, that is, the analytic order, or order of discovery. If, however, the synthetic order, as typified in Euclidean geometry, is used, proceeding from definitions, axioms, and postulates to theorems and problems, "I am, I exist" does not perform the rôle of first principle. When at the request of the authors of the second set of *Objections* Descartes pro-

[3]*Letter to Clerselier,* June or July, 1646, *Œuvres de Descartes,* ed. C. Adam and P. Tannery (Paris, 1897–1913), IV, 444–5.

vided an essay in metaphysics, "drawn up in geometrical fashion," the "I am, I exist" makes no appearance at all. The first proposition asserting existence asserts it of God, and from that Descartes proceeds to the existence of the world. Although he gives a proof for this first proposition asserting existence, he denies that it needs a proof. If his readers only "dwell long and much in contemplation of the nature of the supremely perfect Being . . . from this alone and without any train of reasoning they will learn that God exists, and it will not be less self-evident to them than the fact that the number two is even and number three odd, and similar truths."[4] God's existence, then, could be a first principle of metaphysics, but in so far as the Cartesian method is used in metaphysics the "I am, I exist" will occupy this place, for it is the more easily discovered.

[4]H. R., II, 55.

Index